"Read this book and you will get a deep ı
design is all about. In clear and concise pr
ical work environment can affect the well ___g, ____
iour and productivity of office workers. The book is an excellent blend of
academic insight and practical relevance."

—*Dr Juriaan van Meel, Co-founder BriefBuilder*

"*The Workplace Zoo* is a timely reminder of the need to revise modern organizational practices. Using a mix of applied research and business experience,
Nigel Oseland provides adroit lessons in how to build more meaningful
working lives. The result is a readable, knowledgeable and useable book
that will add to the armoury of the intelligent manager."

—*Dr Craig Knight, Founding Director,*
*Identity Realization Ltd*

"Understanding the impact that buildings have on people is becoming
more important as organisations look to invest intelligently in people and
place. The Workplace Zoo comprehensively equips all interested parties to
navigate the sometimes arduous journey towards an effective environment.
If you have any interest in what lies behind creating a great workplace experience, then this is the book for you."

—*Mark Catchlove, Director, Insight Group –*
*EMEA, Herman Miller*

"Nigel takes us on a journey *Beyond the Workplace Zoo*. It already sounds
very intriguing but is getting more exciting when you began to turn the
pages. He elegantly uses his psychological background and practical experience to explain different theories and how office design can affect our
satisfaction and work performance, and what solutions do work and what
do not. It is not an academic telling but given by a person with long practice. He tells us how to create an office design that you want to and must
return to – socially welcoming, promoting innovation and interaction, and
creating spaces for concentrated work, creativity and relaxation, and more.
As per the title, Nigel tells you how to create an office *Beyond the Workplace
Zoo*. It is a must-read for the designers, architects, and creators of office
space and any manager that wants to ensure that his workforce is effective
and satisfied – full of tips and knowledge that you must not miss."

—*Associate Professor Pawel Wargocki, Interna-*
*tional Centre for Indoor Environment and Energy,*
*Department of Civil Engineering, Technical University*
*of Denmark*

"Nigel Oseland has that rare quality of combining academic rigour with an ability to communicate well on complex subjects. And what could be more complex and timely – and more in need of genuine insight – than the changing nature of work? Nigel's studies over many years make him the ideal person to explore the ways in which people function and interact as animals and the effects on them of the rapidly evolving world of work."

—*Mark Eltringham, publisher of* IN *Magazine and workplaceinsight.net*

"A long-time supporter of our conferences, in *Beyond the Workplace Zoo* Nigel demonstrates one of his key strengths, which is researching and gathering technical information, then distilling and explaining it in a way that we can all understand and apply. The results are key to creating the most modern of workplaces, fit for inhabitation by employees for generations to come."

—*Maggie Procopi, Workplace Trends*

"Nigel Oseland is the Denis Diderot of the office. His history is fascinating; his barbs against architectural determinism and psychobabble, laudable; his solutions, and his accounts of open plan and agile working, commendably balanced; his writing style, terminology and taxonomies, unrivalled in their clarity. In office design, he will persuade many to vote NO – to false claims and false prophets alike. Recommended."

—*James Woudhuysen, Professor of Forecasting and Innovation, Department of Mechanical Engineering and Design, London South Bank University*

"The title of this book (*Beyond the Workplace Zoo: Humanising the Office*) inspires a look under the cover. For those who have often thought of their office as a zoo, this book will illuminate a future that focuses on the fusion of human needs and workplace design. It generates a more innovative approach, focused on value created, rather than the typical emphasis on costs per square foot. The book draws on an extensive body of research in environmental psychology and effectively translates the data and insights into workplace solutions that are better suited to this moment in history that calls for more flexibility and imagination. The proposed new workplaces do not destroy previous efforts, but rather integrate the best features of the of the past with new approaches to technology-aided mobility and the capability to work in multiple places including home. *Beyond the Workplace Zoo* is a valuable and thoughtful look at how learning from the past can illuminate the future."

—*Dr Judith Heerwagen, Affiliate, Faculty of Architecture, University of Washington*

"Nigel's *Beyond the Workplace Zoo* is concise, comprehensive and charming, rich in resource and ideas, a thoroughly worthy addition to the growing body of workplace literature. The particular focus on the need to cater for the subtle yet vital differences between us is timely and important."
—*Neil Usher, Chief Workplace & Change Strategist @ GoSpace AI, Author of* The Elemental Workplace *and* Elemental Change

"Finally! A book that really explains what is happening in the world of workplace. It covers everything from the true value of space to personality profiles, the read is a journey worth taking. It's written in conversation style and addresses the more complicated, academic subjects with practical inter-pretations that can be used in the real world. Possibly the best book on the workplace I've ever read."
—*Paige Hodsman, Concept Development, Workplace Specialist, Saint-Gobain Ecophon*

"Nigel manages to ram the seventy-year evolution of the contemporary corporate office into a new practical and essential guide for all students of the future of workplace. Presented with typical academic rigour, this is a new must read for anyone trying to navigate the complexities of the post-pandemic people + place debate."
—*Tim Oldman, Founder & CEO, Leesman*

"Nigel has written a truly original and insightful book that brings together his academic knowledge in psychology and his vast practical experience in work-place strategy. In this sense the book takes a humanistic perspective on the workplace, explaining how workplaces needs to be designed with peoples' men-tal and physical wellbeing on mind rather than, as is mostly the case, design aesthetics. This book is a 'must have' for anyone concerned with the role of the workplace in supporting individual and organisational performance."
—*Dr Rob Harris, Principal, Ramidus Consulting*

"For too long organisations have been looking down the wrong end of the telescope when it comes to the spaces they provide employees to do their best work. Nigel's experience and expertise are an antidote to that. There hasn't been a conversation on this topic worth having in the last decade that Nigel hasn't been front and centre of so any business would benefit from the thoughts in this work on how evidence backed insights can make a real difference in creating amazing spaces for people to do amazing work."
—*Chris Moriarty, Director of Insight & Engagement, IWFM Institute of Workplace and Facilities Management*

# Beyond the Workplace Zoo

This book begins by outlining the common design mistakes with the modern open plan office and the industry focus on cost that has resulted in the ill-fated *Workplace Zoo*. The requirements of office-based workers according to psychological theory and research are then explained. Dr Oseland references historic studies in psychophysics to describe how to design environmental conditions (acoustics, lighting, temperature, indoor air quality) that enhance performance by supporting basic physiological needs. More contemporary research in environmental psychology investigates how cognition affects our interpretation and response to physical stimuli depending on personality, context, attitude and other personal factors. This in turn informs individual requirements for environmental conditions as well as group needs. Studies in evolutionary psychology and biophilia are also referenced.

The latter part of the book turns to workplace solutions and focuses on how to plan, design and manage offices to accommodate our innate human needs now and in the future. The importance of designing for inclusivity is also recognised, including accommodating cultural, gender and generational differences along with designing spaces for neurodiversity. Dr Oseland's proposed workplace solution the *Landscaped Office* is a revived and revised version of Bürolandschaft with a contemporary twist. The impact of workplace trends, such as agile working and hot-desking, is also explored and found to complement the workplace solution, resulting in the *Agile Landscaped Office*.

This book is key reading for professionals, and post-graduate students, in business, interior design, architecture, surveying, facilities management, building services engineering, HR and organisational or environmental psychology.

**Dr Nigel Oseland** is an environmental psychologist, workplace strategist, change manager, researcher, international speaker, lecturer and published author with 11 years research and 23 years consulting experience. In his consulting business, he draws on his psychology background and his own research to advise occupiers on how to redefine their workstyles and rethink their workplace to create working environments that enhance individual and organisational performance and deliver maximum value. Nigel occasionally lectures at numerous universities on workplace psychology, space management and wellbeing in buildings. He is the programme advisor and a regular speaker at the Workplace Trends series of conferences. Nigel regularly presents at international (academic and corporate) conferences on designing workplaces that enhance performance and wellbeing. He made a TEDx talk in 2018 on How We Can Create Unique Workspaces and introduced the concept of the *Workplace Zoo*.

# Beyond the Workplace Zoo
## Humanising the Office

**Nigel Oseland**

Routledge
Taylor & Francis Group

LONDON AND NEW YORK

First published 2022
by Routledge
2 Park Square, Milton Park, Abingdon, Oxon OX14 4RN

and by Routledge
605 Third Avenue, New York, NY 10158

*Routledge is an imprint of the Taylor & Francis Group, an informa business*

© 2022 Nigel Oseland

*British Library Cataloguing-in-Publication Data*
A catalogue record for this book is available from the British Library

*Library of Congress Cataloging-in-Publication Data*
Names: Oseland, Nigel, author.
Title: Beyond the workplace zoo: humanising the office / Nigel Oseland.
Description: Abingdon, Oxon; New York: Routledge, 2022. |
Includes bibliographical references and index.
Identifiers: LCCN 2021014253 (print) | LCCN 2021014254
(ebook) |Subjects: LCSH: Office layout. | Work environment.
Classification: LCC HF5547.2 .O85 2022 (print) | LCC HF5547.2
(ebook) | DDC 658.2/3—dc23
LC record available at https://lccn.loc.gov/2021014253
LC ebook record available at https://lccn.loc.gov/2021014254

ISBN: 978-0-367-65532-7 (hbk)
ISBN: 978-0-367-65533-4 (pbk)
ISBN: 978-1-003-12997-4 (ebk)

DOI: 10.1201/9781003129974

Typeset in Goudy
by codeMantra

# Contents

*Author biography*                                                    xiii
*Acknowledgements*                                                     xv
*Foreword*                                                            xvi

**Introduction**                                                       1

**1  Introducing the *Workplace Zoo***                                 3
*Origins of the* Workplace Zoo  3
*A human-centric and evidence-based approach  8*
*Legacy workplace issues  11*
*A book of two halves  12*

**PART 1**
**Situation**                                                         15

**2  Fascination with cost**                                          17
*The productivity equation  18*
*The value-cost conundrum  20*
    Property costs  22
    Destination density  24
    Quantifying the benefits  27
*Cost-benefit analysis  34*
*Project evaluation  36*

**3  Psycho-what?**                                          38
*Early psychophysics 38*
*Environmental psychology 41*
   Spaces for people 43
*Evolutionary psychology 46*
   Biophilia 47
   Neuroscience 50
   Anthropology and Dunbar's number 51
*Motivation theory 52*
   Maslow 52
   Herzberg 54
*Personality theory 55*
   Temperaments and psychoanalysis 56
   Trait theory 57
*Sensory processing and multisensory design 64*
*Inclusivity and diversity 66*

**4  The rise and failure of open plan**                     72
*The route to open plan 72*
*Open plan versus private offices 77*
*Benefits of open plan offices 81*


**PART 2**
**Solution**                                                 87

**5  Introducing the *Landscaped Office***                   89
*A (re)emerging workplace solution 89*
*Humanising the office 92*
*Accommodating human needs 95*

**6  It's a jungle in there**                                101
*Evolutionary psychology 102*
*Biophilic design 107*
*Building standards 114*

**7  A plan comes together**                                 116
*'Til desk do us part 117*
*A menagerie of work-settings 119*
   Concentration and confidentiality 120

Collaboration, creativity and connectivity  121
Contemplation and care  124
Core and common spaces  124
*Planning the landscape  124*
Workplace layout  124
Workspace zones  125
Work-settings  126
*Bringing it all together  128*
Look and feel  129
Reckoning occupational density  130

8  **Choice, sharing and agile working**                                     132
*Types of agile working environments  133*
*Benefits and barriers to agile working  137*
Advantages of agile working  137
Challenges of agile working  139
Work Patterns  140
*Adoption of agile working  142*

9  **The great indoors**                                                     145
*Acoustics and noise  146*
Acoustic standards  146
Sound versus noise  147
Solutions for noise  147
*Thermal comfort  155*
Temperature and performance  155
Thermal comfort standards  156
Adaptive comfort  161
Solutions for thermal comfort  163
*Indoor air quality  166*
Indoor Air Quality regulation  166
IAQ and performance  167
Scents and odours  169
Solutions for air quality  170
*Lighting and daylight  170*
Lighting and performance  170
Lighting standards  171
Lighting and preference  173
Daylight and colour spectrum  174
Solutions for lighting  176
*Design for all not just the average  177*

**Epilogue**           179

**10 Concluding remarks on the *Landscaped Office***      181
    *Part 1: Situation* – Workplace Zoo  *181*
    *Part 2: Solution* – Landscaped Office  *184*
    *Implementing the* Landscaped Office  *187*
    *Beyond this book*  *188*

    *References*          191
    *Index*          205

# Author biography

Dr Nigel Oseland is an environmental psychologist, workplace strategist, change manager, researcher, international speaker and published author with 11 years research and 23 years workplace consulting experience.

Nigel draws on his psychology background and his own research to advise occupiers on how to redefine their workstyles and rethink their workspace to create working environments that enhance individual and organisational performance, delivering maximum value and benefit. He specialises in advising his clients on how to develop and implement workplaces that meet psychological needs and facilitate collaboration, creativity and concentration. Nigel has advised corporate businesses, public-sector bodies and educational institutions in the UK and throughout EMEA.

Before setting up his own workplace consulting practice, Nigel was a researcher at the Building Research Establishment and completed his doctorate with the School of Engineering at Cranfield University. He then went on to become a workplace consultant at Johnson Controls (within their Facilities Management business), Swanke Hayden Connell Architects and specialist workplace practices AMA Alexi Marmot Associates and DEGW. When it comes to workplace design and planning, Nigel has an all-round experience.

Nigel is also an active speaker, blogger, workplace pundit and university lecturer. His current topics of interest include psychological needs, psychoacoustics, productivity, personality factors, remote working, collaboration, creativity, wellbeing, biophilic design and post-occupancy evaluation. He is the programme advisor for the Workplace Trends series of international conferences.

Nigel broadcasts weekly on Radio Dacorum, his local community radio station. He is a Distinguished Toastmaster and a long-standing member of Berkhamsted Speakers Club. Nigel set up and part-owned his town's brewery, but is now focussing on training to become a beer sommelier; he frequently organises beer festivals, on-line tastings and other related events. In his leisure time, Nigel regularly walks through the Chiltern countryside, frequently partakes in downhill mountain biking and occasionally scuba dives in warm waters.

# Acknowledgements

I may be the sole author of this book, but I have received many contributions and much encouragement along the way. I am indebted to my esteemed peers and reviewers for their support and recommendations, which I have attempted to include, so thank you Neil Usher, Pawel Wargocki, Rob Harris, Juriaan van Meel, Peter Andrew, Craig Knight, Paige Hodsman, Matt Tucker and Mark Eltringham. I am also grateful for comments, images and other useful bits and pieces received from Mark Catchlove, Duncan Young, Nick Pell, Sally Augustin, Iain Smith, Tim Fox, Chris Parker, Yuri Martens, Hermen van Ree, Nigel Bunclark, James Woudhuysen and Judith Heerwagen. I must give a special mention and thank you to Fiona Duggan for all her suggestions, support and redrawing the images; Melanie Thompson for her expert editing and restructuring; and Robin Dunbar for his kind words. Last, but by no means least, a big, huge thanks to Maggie Procopi ("the wife") for encouraging me to write a book and for her enduring reassurance, inspiration and tolerance.

# Foreword

In our search for efficiency and cost savings, we often seem to lose sight of the fact that the workplace is a social environment. Every business and every organisation is a social microworld whose success depends not just on the health and motivation of its workforce but to an even greater extent on how well they get on socially with each other. The world of work is a social world, and everything from the flow of ideas and creative insights to the small favours and obligations that enable things to get done depends on human relationships. Human relationships depend not on management systems and financial spreadsheets but on the simple everyday processes that characterise the world of family and friends – the casual environment of the pub rather than the world of business strategy and corporate mission statements.

How we design the physical space within which this social world functions can dramatically affect the efficiency and productivity of the organisation that inhabits that space, be it a school, a hospital, a government department, a factory or a business. They all work, or don't work, because they are microcosms of the human society that ebbs and flows outside their doors. When we ignore that and design buildings and their interiors by purely functional criteria and the demands of price, we jeopardise the success of the occupants before they have even put their feet under their desks.

The psychological dimensions of the workplace can sometimes be very subtle. Even the size of the task groups that work most efficiently derives from deep down inside minds that evolved during the course of several million years on the savannahs of Africa, honed in more recent history in the residential suburbs of metropolises, rather than the artificial environments of Wall Street and Whitehall. This psychology has its expression in Dunbar's Numbers, the series of group sizes at around 5, 15, 50 and 150 that work most efficiently together. The optimal number of people for a job depends not on the size of your room but on how closely the group has to work together and how intimately its members need to understand each other's thought processes for the particular task in hand.

Designing the space and layout of an office in ways that map onto these numbers can dramatically influence the social psychology and group dynamics of the shop floor. Get it wrong, and you torpedo the success of an organisation even before you've handed over the keys. In a word, the structure of a business is not something to shoehorn into whatever space you can afford, but rather the space has to be built around the structure of the business. The task of the architect and the interior designer is not the design of physical beauty so much as the design of spaces that facilitate the meeting of minds and the social flow of the human relationships that underpin the success of any organisation.

In this elegant book, Nigel Oseland distils a lifetime of practical experience in the design of workspaces bolstered by the insights of an active research career to provide us with both an exquisitely clear guide to best practice and an explanation of the science behind it.

Robin Dunbar
University of Oxford

# Introduction

# 1   Introducing the *Workplace Zoo*

After 30 years of researching and advising on the impact of the built environment on comfort and performance, after over 20 years since publishing my last book, after writing over 100 related articles, after several years of blogging, and after attending, chairing and speaking at numerous conferences on the subject, I felt it was time to capture my thoughts and advice on how to design workplaces, in particular offices, in one short, easily digestible volume.

Work is continually evolving, but there are basic and innate human needs that must be met in order to ensure the wellbeing and performance of each worker. In the modern office, these basic human needs are often overlooked or even ignored in favour of a reduction in space and associated property cost savings, thus creating the ineffective and unpopular *Workplace Zoo*. This book is a timely reminder of how to create workplaces that enhance the wellbeing and performance of a variety of individuals. It builds on long-standing research and existing best practice design, that is often forgotten or simply ignored, reinterpreted with a contemporary twist.

## Origins of the *Workplace Zoo*

Before offering my advice, I need to explain further the phrase *Workplace Zoo* used in the title of this book and throughout. There are several key sources that influenced the title. As a 16-year-old budding social scientist, I was fascinated by the writing and TV programmes of Desmond Morris, a zoologist and people watcher. Back in 1967, Morris wrote *The Naked Ape* in which he claimed that a lot of modern behaviour harks back to the era when humans evolved from primates into prehistoric man, hence hairless or naked apes. He explained how grooming, sexual pairing, monogamy, prolonged development of the young (neoteny) and other factors increased social cohesion, group collaboration, language, intelligence and, ultimately, the success of the human species.

DOI: 10.1201/9781003129974-2

In his second book, *The Human Zoo*, Morris compares the behaviours of human city dwellers to the inhabitants of zoos. He refrains from using the term "concrete jungle" to describe the city, as "jungle" infers a natural environment whereas a "zoo" is artificial. Morris believes that undesirable behaviours (such as violence towards fellow humans, self-mutilation and eating disorders) are related to the stress of living in over-populated, crowded habitats and from boredom due to lack of stimulation associated with easy modern living. He points out that such behaviours are more associated with captive animals living in the confined cages of 18th- and 19th-century zoos (Figure 1.1) rather than those roaming free in their natural environment, in the jungle or elsewhere. Morris might regard the modern office as a form of human zoo, but I wanted to distinguish between the macro level of the city and the relatively micro level of the office, hence *Workplace Zoo*.

I am a fan of the modern zoo. Many years ago, I visited Colchester Zoo with my children, and I came away extremely impressed with the quality of the animal enclosures. Clearly, a lot of thought had gone into their design and what was most impressive was how each enclosure was clearly designed to meet the specific needs of the occupying species. A great deal of effort was made in simulating the animals' natural environment and ensuring that they were comfortable. This was evident in the way that the different animals behaved (none of the undesirable behaviours noted by Morris: no pacing or listlessness) and through the success of their breeding programmes.

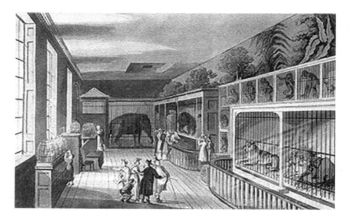

*Figure 1.1* Traditional early 19th-century zoo (Unknown author, public domain, via Wikimedia Commons).

After my zoo trip, I returned to the office, I was working for a large corporate at the time, and it struck me how homogeneously designed each floor of our office building was despite having quite different occupants. I acknowledge that humans (*homo sapiens*) are all the same species, but we are very different "animals". We have a shared anatomy and physiology, but nevertheless, individuals differ. For instance, hearing, sight and physique vary considerably among the population, and between those working for the same organisation. Physiology is affected by age and health, but other personal and demographic factors (e.g. size, gender and ethnicity) and psychological factors (e.g. attitudes and mood) will also influence our preferences and needs. Research, including my own, shows that personality traits affect our ideal environmental conditions and preferred workspaces. Situational factors such as job role, and the associated work activities being conducted, will also vary and affect our workplace preferences.

So, I started wondering whether any lessons learned in modern zoo design (Figure 1.2) are relevant to the workplace. It turned out that I was not the first to make this comparison. Psychologist Judith Heerwagen (2008) suggests:

> For insights, it is useful to look not at buildings, but at zoos. Zoo design has gone through a radical transformation in the past several decades. Cages have been replaced by natural habitats and geographic clustering of animals. And, as in nature, the animals have much greater control over their behaviour. They can be on view if they want, or out of sight. They forage, play, rest, mate, and act like normal animals … A key factor was concern over the animals' psychological and social wellbeing. Zoos could keep animals alive, but they couldn't make them flourish.

Heerwagen proposed that we learn from the new philosophy of enriched zoo enclosures, which provide for wellbeing rather than simple survival. While controversy still surrounds keeping animals in captivity, the better zoos have successful breeding programmes and have reared endangered species in captivity.

So it appears that we can learn from the basic design principles of zoo enclosures. For example, zoos would not place different animals in the same enclosure unless the environments suited them all, and there is a symbiotic relationship between those animals. Our different human "animals" are often placed in the same environment regardless of the specific needs required to allow them to thrive.

Provision of a suitable environment is the most fundamental of key principles in zoo practice. The UK Government recommends that "the temperature, ventilation, lighting and noise levels of enclosures must be suitable for the comfort and wellbeing of the particular species of animal at all times" (DETR, 2000).

*Figure 1.2* Modern zoo enclosure (Kevin1086, CC BY-SA 3.0, via Wikimedia Commons).

Painstaking effort has been taken, with meticulous attention to detail, to ensure the enclosures in modern zoos provide each species and sub-species of animal with the best environment to allow them to flourish. But although some effort is made to ensure that comfortable environments are provided in the workplace, post-occupancy evaluations (POEs) repeatedly show that occupants have low levels of satisfaction with temperature, air quality and noise. Our individual preferences, behaviours and activities mean that it is a challenge to designers and engineers to provide comfort for everyone, but such a repeatedly low level of satisfaction is neither acceptable nor good workplace design.

Regarding acceptable temperatures for animals, it is quite common for temperatures in London Underground (tube) carriages to reach 40°C in summer. When I was unfortunate to be standing and sweating in a crowded carriage, I often wondered to myself why it is illegal to transport livestock at temperatures above 35°C (DETR, 2010) but not humans. I appreciate that the animals have no choice in the matter and their journey is longer, but I am not sure that those humans, dependent on the tube, have that much choice either. Fortunately, the London Underground has since introduced cooling into some of its carriages, and many trains are now air-conditioned, but pre-pandemic the systems continued to struggle at peak times due to over-crowding. Similarly, office environmental conditions, such as temperature, air quality and noise, are all adversely affected by overcrowding due to in-creased occupational density.

It could be argued that zoo enclosure design is easier than workplace design because it accommodates a single species with a basic animalistic drive for survival. Ethnic groups across the world have adapted to different climates, including some evolved physiological differences, but neverthe-less, we are one species. However, as Richard Dawkins postulated in *The Selfish Gene* (1976), the single motivator for human behaviour is survival. So, it could be counter-argued that the design of both zoo enclosures and workplaces comes down to a thorough understanding of the range of occu-pants' needs and designing to meet them. There are many factors that drive how humans behave daily. We are a complicated species, separated from other animals by our intelligence and personality, as well as neocortex size and opposable thumbs and so on, and these individual differences lead to certain behaviours and needs. In a zoo, if an animal exhibits a particular characteristic that requires a specific environmental adjustment for them to flourish, then it is very likely that the zookeeper would make that provision rather than ignore it or assume an environment with middling conditions will suit all. However, this is not the case in the workplace; we often provide a homogenous environment for a "single species" and there is little recogni-tion of individual differences and the associated requirements to enhance comfort, wellbeing and performance.

Relating the workplace to a zoo has been used before, but not in the context of office design. Motivational speaker Nigel Risner published *It's a Zoo Around Here* (2003), in which he explains how to identify your communication style and that of others so you can communicate more effectively with them. The communication styles are linked to the dif-ferent dominant traits of office workers. Risner illustrates four dominant communication styles using animals: dolphin, monkey, lion and elephant. In his later book (Risner, 2020), he remarks that "the workplace is a bit of a zoo and it's often a pretty emotional, chaotic one at that. Moreover, a zoo isn't a natural environment for any of the animals who live in it. The workplace isn't any different". But his focus is still on communication rather than design. Similarly, Karin Stangl's (2017) book of poems, called *The Workplace Zoo*, illustrates familiar office worker stereotypes using an-imals. Apologies to Stangl, but I am repurposing the phrase *Workplace Zoo* to illustrate how modern office spaces that are poorly designed and poorly managed hark back to the days of cage-based zoos. In a nutshell, I propose that:

The *Workplace Zoo* is the modern interpretation of the open plan of-fice. A high density, overcrowded space with serried rows of desks, few partitions, limited facilities and poor environmental conditions. The focus is foremost on space and cost efficiency rather than designing and planning for human needs to enhance the wellbeing and performance of its occupants.

## A human-centric and evidence-based approach

Many years ago, Le Corbusier (1927) claimed "a house is a machine for living in" so logically it follows that "an office is a machine for working in", or as a previous co-author of mine eloquently put it "buildings used as an office have a simple purpose: to enhance the performance of the organisation using them" (Oseland and Bartlett, 1999). My starting point is therefore designing offices for the individuals that make up those organisations, and consequently, this book offers a human-centric approach to design.

There is a subtle difference between human-centric and human-centred design. Both approaches improve satisfaction and wellbeing whilst counteracting any adverse effects on health, safety and performance (*ISO 9241-210*). Human-centric simply refers to a focus on human needs and designing solutions (possibly a system, technology or building) to meet those needs, whereas human-centred design focusses on end-user needs but also involves end-users in all steps of the design process. Personally, I believe that human-centric implies that the solutions – in our case, office designs – may be tailored to individual needs or at least to the needs of a small group or an organisation.

Recognising and designing for specific and individual needs, rather than assuming there is a single generalisable office design solution for everyone, is the crux of this book.

My chosen profession is workplace consulting. It mostly entails understanding organisational needs, recommending the appropriate workplace strategy to support those needs, developing the design brief and assisting in implementing the new workplace through a structured change management process. Whilst I am not an architect, interior designer, engineer or surveyor, I am a practising environmental psychologist with a keen interest in both people and buildings. Environmental psychology is basically the scientific study of the human mind and behaviour in the built, and natural, environment. I love my discipline of psychology, but I do not appreciate psychobabble, pseudo-psychology or pop-psychology. As a practising psychologist and researcher, my advice is grounded in solid evidence (i.e. evidence-based design).

Evidence-based design is often defined as the process of constructing the best physical environment based on scientific research, and Hamilton and Watkins (2009) propose that the evidence may come from a range of sources including research and practice. Fundamentally, evidence-based design builds on preceding knowledge and adopts what is demonstrated to work well, rather than simply rely on intuition, hearsay or anecdotes. I am in favour of evidence-based design not just because of my research and psychology background but also because of disappointing experiences on design projects. I recall one architectural design team recommending placing a large unenclosed 12-person meeting space in the middle of my client's open plan office. I questioned their reasoning, explaining that the occupants were assessors and mostly involved in focused tasks requiring concentration. Their response was "it works for us in our studio". I was amazed at such a naïve answer – why would they believe that an environment that suits one group of people would be appropriate and acceptable for another group carrying out different activities in a different organisation with different circumstances, different demographics and different personal factors like age and personality?

The architects' reasoning for an open meeting space was just one up in the league of poor design excuses to the more common response of "It will look cool". I appreciate that there are different success criteria for office buildings. Such as Vitruvius's three principles of architecture, dating back to 30–15 BC but still cited by architects, stating that a building must exhibit *firmitas, utilitas* and *venustas,* that is, be solid, useful and beautiful (Figure 1.3). So, yes, a building should be appealing, aesthetic and attractive, but first, it must be functional and structurally sound. The mantra "form follows function", coined by architect Louis H Sullivan (1896), is a useful reminder of Vitruvian principles. Buildings are often referred to as iconic, and a large corporate occupier might request that their new office building is striking, stands out, has presence and makes a statement (with

*Figure 1.3* Vitruvian architectural principles.

some external branding it may even act as a permanent ongoing advert) sometimes forgetting that it also needs to work *as a building*. The workplace consulting practice DEGW (Allen et al., 2004) refers to a building that represents the occupier's culture and brand as "expression" – one of their "Three Es" with the other two being "efficiency" and "effectiveness".

I recall hearing a story, which I hope is true, that when Swiss Re commissioned a new building in London their brief to the architects, Fosters + Partners, was to make it so iconic that when visitors arriving at Heathrow Airport mentioned the Swiss Re building to their taxi driver, the driver would not only recognise the occupier's building but know its location. The building did indeed become famous, but it became known by its nickname derived from its shape, *The Gherkin*, rather than by its occupier – the building has since been rebranded *30 St Mary's Axe* but locals still refer to it as *The Gherkin*. It is easily argued that *The Gherkin* has *firmitas* (it has a complex and interesting structure) and *venustas* (there is beauty, it stands out on the skyline and is striking), but it perhaps lacks *utilitas*. Case in point, a colleague of mine drawing up space plans to entice new tenants, discovered that tapering circular floors are difficult to efficiently space plan, resulting in unusable and wasted expensive office space. Swiss Re sublet the office floors but it took a while for some of those floors to be leased, and this may have been partly due to their poor space efficiency. The top dome of *The Gherkin* was used for events and presentations. I made a presentation there, and whilst the space offered impressive panoramic views over London, the acoustics were terrible because of the reflective and reverberant nature of the dome, and the presentation screen was washed out due to the excessive (normally appreciated) daylight – making presentations uncomfortable for both the speaker and audience. In fairness, the dome was probably not initially intended for use as a presentation suite, and a retrofit solution was later implemented.

It is true that buildings are also investments, favoured by pension funds, with many shares tied up in property with long-term lucrative returns. If a building is an investment, then the focus is on maximising the return by minimising costs, increasing the lettable space and leasing/selling it quickly. Chris Kane (2020), ex-head of Corporate Real Estate (CRE) at Disney and the BBC, points out that "for many in the industry, it is still about 'the deal'. This short-term, 'let's make big bucks' view, coupled with the clichéd developers' attitude of 'build it and they will come', is a very aggressive standpoint". The developer and investors may make savings by compromising on *firmitas*, *utilitas* and *venustas*, but ultimately, this is not sustainable because potential occupiers will choose to opt for better, more suitable, buildings on the market. A human-centric and evidence-based approach informs successful design that is attractive to future tenants.

An evidence-based approach should not stifle innovation: the evidence is open to different interpretation and implementation. Nevertheless, intuitive and innovative approaches should be evaluated before being adopted wholesale. With this in mind, throughout this book, I refer to published and unpublished case studies along with my own and my peer groups' project experience, but I will mostly rely on the psychological literature. Fortunately, there is a rich body of psychological research and literature on human needs. Unfortunately, much of it is complex and contradictory and has either been misunderstood, forgotten or simply ignored. This book aims, in part, to demystify and simplify those hefty academic tomes and convert them into pragmatic advice.

## Legacy workplace issues

I wrote this book during the Covid-19 pandemic and resulting lockdown, which caused organisations to close their offices and force their employees to work from home. Despite that, this book is not intended as a guide on how to design offices specifically for the post-pandemic era, especially as other similar outbreaks are likely to occur in the future. Nonetheless, at the most basic level of human needs, offices (like zoos) should not harm the health or safety of their occupants, so this book includes generic recommendations on creating more healthy working environments. A few major historic health risks in offices due to poor air quality are highlighted below.

Covid-19 is not the first health crisis that has resulted in the closure of offices. Previous Corona viruses, MERS-Cov and SARS, have also disrupted office work over the last 20 years but not on such a large and widespread scale as Covid-19. Furthermore, the track record of the modern office causing ill-health goes back quite a while.

I associate Legionnaire's disease with the early 80s, but I was surprised to find that there are still cases today – some 30 in the last 10 years. Legionnaire's can be fatal and is primarily caused by aerobic bacteria contaminating a building's water system. It is more common in hotels and, ironically, hospitals but occasionally is found in working environments.

In my early days, as a Government researcher in the late 80s, I was part of a European team investigating Sick Building Syndrome (SBS). It is identified by a range of symptoms, such as headaches, dry eyes, nose and throat irritation and fatigue, that result in poor performance and absenteeism. It is not at all clear of the cause of SBS, but it is most likely due to reduced ventilation rates and increased recirculated air in order to save energy and money. This lowers the fresh air intake and in turn allows the build-up of pollutants, dust and mould. SBS is also associated with flickering fluorescent lighting and was found to be more common in workers feeling stressed by their workload.

The practice of increasing recirculation, and lowering fresh air, also escalates the risk of cross-infection of airborne diseases. Ventilation rates in offices are based on occupancy levels such that increasing the occupational density, by cramming in more desks, without increasing the ventilation will exasperate any potential health risk. Over the years, densities have gradually increased in offices, see BCO (2018), primarily to save costs, but in return, it has placed strain on the ventilation system, temperature control, acoustics, accessibility and facilities. It is well-documented that, in the animal kingdom, overpopulation and overcrowding of habitats often lead to disease and higher mortality rates (Herman, 1969). Ethologist John B. Calhoun (1962) found that rats in overpopulated nests suffered from abnormal and destructive behaviours resulting in high mortality rates, despite having unlimited food and being protected from predators. A return to lower density office spaces, which better support health and performance, is long overdue (and covered in more detail in the following chapters).

It is not just airborne diseases that can cause ill-health in offices. Charles Gerba (2002) and colleagues at the University of Arizona found that desks contain nearly 400 times as many microbes than toilets. The offices studied contained approximately 135,000 germs per square centimetre, predominantly on keyboards, telephones and mice, compared to 300 germs per square centimetre for toilets. At the time of the study, one microbiologist commented that it was impossible to create sterile offices, but the risk of infection can be minimised by washing hands regularly and using alcoholic wipes on office furniture. Advice even more applicable in the era of Coronavirus.

Behaviour can also escalate threats to health. Presenteeism is the practice of coming into the office even when feeling ill. The pressures of work commitments, perhaps terms and conditions, over-enthusiasm, not wanting to "miss out" or be forgotten, may mean that staff turn up to the office spreading disease rather than take time out from work to recover.

The main issue, it seems, with lowering density, increasing fresh air ventilation and more regular cleaning, is the additional cost to the occupying organisation.

## A book of two halves

This book has two main parts: the situation and the solution. The first half (Chapters 2–4) mostly focusses on basic human needs and how they relate to the requirements in the office that are necessary to enhance our wellbeing and performance. The latter half of the book (Chapters 5–9) provides my proposed solutions relating to the design and management of offices.

The proposed office solutions are based on human needs, so they are not time-bound and as such are relevant to the post-pandemic workplace.

- *First part: Situation* – The primary purpose of an office is to facilitate the business and activities of the occupying organisation. For an organisation to be successful, for it to perform well, it requires the individuals within its workforce to be performing to their maximum potential. The workplace design, layout, facilities and operation must therefore support the wellbeing and performance of the staff. However, the CRE industry[1] and to some extent the leadership team of organisations occupying office buildings often view the workplace as a cost burden rather than an enabler or investment in people. This is confounded by a difficulty in measuring performance compared to measuring cost. The focus is therefore often on managing costs which constrains the workplace design and hinders providing the best facilities for the workforce.

  A misdirected approach to office design and planning has, in many cases, resulted in the *Workplace Zoo*, a modern poor representation of the open plan office concept. Now is the time to remodel the office to address our psychological and physiological needs. Office design needs to be evidence-based and human-centric to ensure that it meets those basic needs rather than simply ignore them or have a detrimental effect. The psychological and related literature provides ample evidence on how to design offices to meet our innate and evolving human needs. The design should allow office inhabitants to flourish and thrive rather than simply survive.

  Office design is often complex and to be successful requires a truly transdisciplinary approach, it requires a concerted input and effort from multiple disciplines. The first section of this book covers human needs and the impact on workplace design. As a psychologist, I draw heavily on the fields of psychophysics, environmental psychology, evolutionary psychology, personality theory and motivation theory. However, I also draw on relevant research in the related fields of physiology, sensory design, anthropology, sociology, philosophy, neuroscience, economics, architecture, inclusivity and zoology.

- *Second part: Solution* – Research evidence indicates that a shift from being predominantly cost-focussed to human-centric in workplace design will undoubtedly improve our wellbeing and performance. I have provided the foundation for an improved working environment, which is my updated version of the *Landscaped Office*. My focus in this book is the office, but basic human requirements are needed regardless of the location that work takes place.

The *Landscaped Office* is not an all-encompassing reinvention of the workplace; rather, it is a reminder of long-standing but often forgotten or ignored research and a return to the *Bürolandschaft* space planning concept, somewhat reinterpreted with a contemporary twist. It is a viable and much-needed alternative to the repeatedly found homogenous, uninspiring, unhealthy and unproductive offices encapsulated in the *Workplace Zoo*.

There are four fundamental physical components to the *Landscaped Office*: workspace layout, work-settings, interior design and environmental conditions. The proposed solutions are informed by the individual human needs highlighted in the first half of the book. Furthermore, the *Landscaped Office* is complemented by agile working practices, underpinned by enabling technology and an appropriate organisational culture.

The final chapter provides a summary and reflects on the way forward along with guidance on the adoption and implementation of the *Landscaped Office*.

## Note

1 Throughout this book, I use the term "Corporate Real Estate (CRE) Industry" as a broad catchall phrase for those disciplines focussed on office design, acquisition, build, ownership and operation. This includes commercial real estate, facilities management, property advisors, property investment, project management, quantity surveying, architecture, interior design, workplace strategists and other relevant practices. The CRE industry includes both in-house teams and external advisors. I also use the term "Workplace Industry" as shorthand for those involved in the briefing, planning and designing of office spaces, predominantly workplace consultants, interior designers and change managers supported by HR and IT.

# Part 1
# Situation

# 2 Fascination with cost

For many years, leading lights in the CRE industry including Paul Morrell, the first UK Government Chief Construction Adviser, have referred to the connection between workplace design and productivity as the "search for the Holy Grail".[1] There is a view that the relationship is intangible and that the benefits, or the negative impacts, on the workplace are not demonstrable. However, I consider this view to be mainly due to wrongly perceived measurement difficulty and incorrectly perceived lack of evidence.

One consequence of such widespread beliefs is that the impact of the workplace (space, environmental conditions, facilities, etc.) is often ignored, excluded or forgotten in both the design process and the business case. As measuring productivity is deemed too difficult, for most workplace fit-out and refurbishment projects, the business case is simply weighted in terms of reducing the more tangible property costs. Such an approach promotes the workplace as a cost burden (on one side of the cost-benefit analysis) rather than a potentially lucrative return on investment for the organisation. Kane (2020) proposes that, rather than focus on cost, those in corporate real estate and facilities management actually "have a role to play in helping business leaders understand the link of how a corporate property portfolio delivers better business performance through better workplace performance, therefore maximising the return on investment".

Moreover, in completed projects, the effect of office design on the occupants is not thoroughly tested, documented or shared across the workplace community. So, we may never fully understand whether the design had a beneficial or negative impact on the organisation. So long as the new office design reduces property costs, with no blatant impact on the business, it will be deemed as successful, regardless of the unmonitored consequences to the business. This perpetuates further the notion that the workplace is a

DOI: 10.1201/9781003129974-4

cost burden. Zoos are more concerned with animal welfare than productivity *per se*, but nevertheless, they evaluate the impact of a new space on its inhabitants. Many zoos monitor animals in their new enclosures including regular observation of behaviour comparing it to that of the wild (using activity logs and ethograms), measurements of cortisol to assess stress levels, records of disease prevalence and reproductive success.

Organisations such as the European Association of Zoos and Aquaria (EAZA, 2013) collate and share guidelines and lessons learned from their member zoos. In contrast, case studies and occasional research on the impact of the workplace on productivity exist, but there is no official central industry repository of lessons learned for designers to draw upon, although the *Leesman Index* has made an admirable start (Leesman, 2019a, 2019b). The evidence can also be contradictory or so research-focussed, so academic, that it fails to present any easily assimilated pragmatic guidance.

So, it is understandable that, historically, much of the CRE industry tends to use cost alone in the business case for new workplace projects. While it is understandable, it is nevertheless an ill-advised and flawed approach to workplace strategy and design. To better understand the relationship between workplace design and productivity, the Holy Grail, it is first necessary to understand what productivity actually means in context of the CRE industry.

## The productivity equation

A well-designed and managed workplace offers many benefits to the occupying organisation, but it is productivity that most businesses are interested in. Notably, the Institute of Workplace and Facilities Management's *Stoddart Review* "reveals that an effective workplace can improve business productivity by as much as 3.5%" and "Economist Duncan Weldon believes that could add up to £70 billion to the UK economy"[2] (BIFM, 2016).

It is worthwhile noting that the terms "performance" and "productivity" are often intermingled, but there is a subtle difference. There are many definitions of productivity, but my preferred one is that used by most productivity researchers (for review, see Oseland, 1999) who express it in terms of efficiency and a rudimentary ratio:

Productivity = Output ÷ Input

In business, productivity is often measured as a deliverable, product or service (the output) as a ratio of the time, effort and cost of producing it (the input). To increase the productivity, businesses look to maximise staff performance, whilst minimising the time, effort and other costs to the business.

As a core focus of the CRE industry is the workspace and the associated cost of providing it, a more appropriate, albeit simple, productivity equation might be the ratio of staff performance to property costs, see Figure 2.1. This ties the workplace to the wider business goals and objectives rather than it being viewed as a stand-alone necessary cost burden. This is akin to two of the "Three Es" of DEGW (Allen et al., 2004) with performance equivalent to "effectiveness" and cost relating to "efficiency" (the third E is "expression"). Incidentally, the term "space efficiency" technically means achieving most benefit from the minimal space. Therefore, like productivity, efficiency is also a ratio, but some view space efficiency as simply being a reduction without considering the impact on benefits.

Paul Bartlett and I (1999) illustrated the relationship between cost, performance and the impact on productivity, as shown in Figure 2.2. We proposed that, to maximise productivity, the CRE industry should seek to increase staff performance whilst reducing the property costs. In contrast, cutting costs resulting in lower or negative performance diminishes productivity. Reducing costs whilst maintaining performance is also productive to some extent, whereas high costs resulting in low performance is the worst-case scenario.

A poor workplace may not only incur additional staff costs, perhaps through recruitment and training, but will also directly impact on staff

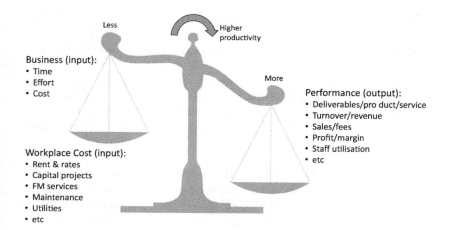

*Figure 2.1* Balancing cost and performance.

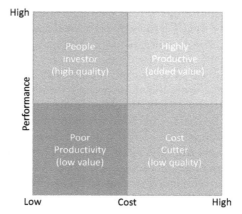

*Figure 2.2* The relationship of cost and performance on productivity (Courtesy of Oseland and Bartlett, 1999).

performance. Not monitoring or, even worse, ignoring the effect of the workplace on the workforces' performance demonstrates poor business acumen. To truly measure productivity involves measuring and monitoring performance as well as costs.

## The value-cost conundrum

Over many years, much of the CRE industry, along with the leadership team of office occupiers, have cultivated an unhealthy skewed fixation with reducing property costs. As, CRE advisor, Chris Kane (2020) points out:

> The last decade has seen incessant effort on driving down occupancy costs and it has been particularly relentless in FM. However, from 2015 onwards, industry commentators ... started a refocussing exercise looking into real estate as a driver of value not just an overhead cost.

He describes a laudable exercise by the enlightened, but it is still not common practice, and the focus on reducing cost to maximise the return in investment applies to the wider CRE industry and the occupier as well as facilities management (FM). Kane goes on to state that "old habits die hard, especially if they are aligned with vested interests and profits in both the business and property sectors".

Of course, it is a perfectly sound business sense to attempt to save money on the construction, operation and management of an office building, but it is only good sense if it does not adversely affect its workforce or impact on the success of the organisation. Tim Oldman of the *Leesman Index* notoriously reported that having surveyed "155,000+ employees worldwide, the latest Leesman figures (Q1, 2016) have revealed that only 55% of employees believe their office environment allows them to work effectively" (Oldman, 2016). Clearly, the approach to workplace design and management is not quite working!

Most industries assess value for money when evaluating proposals, goods, services and products. The decision to go ahead is based on what is being offered for a set cost[3] rather than the cost alone and usually a cost-benefit analysis is included in the business case. For example, there are clear design guidelines for new zoo enclosures setting out what must be provided to "meet the physiological and psychological needs of the animal" (DEFRA, 2012). Such requirements will be included in the business case for the enclosure. Similarly, a new workplace will need to meet a range of building standards, but such requirements might be considered the minimum. The impact of the proposed workplace on the occupants' comfort, wellbeing and performance is usually discussed, but it is rarely quantified and included in a cost-benefit analysis. In many cases, trust is placed in the experience of the designers, and it is simply assumed that the new workspace will meet the requirements of the occupants, enhancing their wellbeing and performance. Therefore, cost is often the predominant criteria for informing property decisions at the business case stage and it dominates other stages of the building lifecycle, including the operations stage.

For an organisation to get the most out of its employees, it will provide them with the best technology, training, business processes, management, etc., in other words the organisation provides an organisational infrastructure that supports its own needs. Yet, the workplace is also part of that infrastructure, and it's a tool for the job and should be treated so. We should consider workplace projects in terms of the return on investment of staff rather than as a cost burden to the organisation. Chris Kane (2020) calls for a more "people-centric" approach suggesting that "Rather than concentrating on space standards, fitting in more people per square foot and cost efficiency, the emphasis should shift on how the workplace can become an effective tool in enabling people to work in the best way possible". In other words, the value offered by the workplace should be considered and not just its cost and that requires a comprehensive cost-benefit analysis.

Predominantly concentrating on saving costs without a full understanding of the consequences demonstrates poor business acumen. The CRE industry, and occupier, focus on cost is due to several reasons including (i) the cost of property, (ii) the legacy of space efficiency and density, and (iii) the difficulty in quantifying the benefits compared to costs.

### Property costs

Office rents have been fairly stable over the last decade or so and in-line with inflation. Nevertheless, for most organisations, property costs are one of the biggest expenses with an initial outlay in capital expenditure, for designing and fitting out the property, followed by regular operational expenditure due to a legally binding commitment to pay rent. The financial commitment varies depending on lease lengths; it could be a long-term commitment with lower rent or perhaps a short-term commitment with higher rent. Breaking the lease early usually incurs heavy financial penalties.

The key asset of any organisation is, of course, its people, and they also happen to be the most expensive element. Consensus is that the second biggest cost to an organisation after its employees is its property, depending on the type of organisation and the level of remote working. Many authors have illustrated the ratio of staff to property costs using pie charts to show that salaries are around 85% of the total business cost over the life cycle of an office (see Figure 2.3). As far back as 1965, the National Bureau of Standards estimated employee costs to be approximately 86% with property costs at 14%, and the British Council for Offices (BCO) reported similar findings 40 years later (CABE and BCO, 2006). More recently, the World Green Building Council estimated staff costs to be closer to 90% with approximately 9% property/rental costs and the remaining 1% being energy costs (Alker, 2014). A more thorough report by the BCO (Harris and Hawkeswood, 2016) found that although property costs are indeed around 15% of business costs, the staff costs are closer to 55% with the remaining 30% made up of other business costs; nevertheless, property costs are high.

Productivity researchers often use staff costs to quantify and monetise any changes in performance. To illustrate, a 5% increase in performance by the same number of employees is equivalent to saving 5% in staff costs. Based on the 85:15 staff:property ratio, several researchers have proposed that an increase in performance of approximately 15% would cover the cost of the property by off-setting the equivalent in employee costs, see Oseland (1999) for a review. However, the 15% is likely to be an under estimation because, in many corporate businesses, the employees are expected to

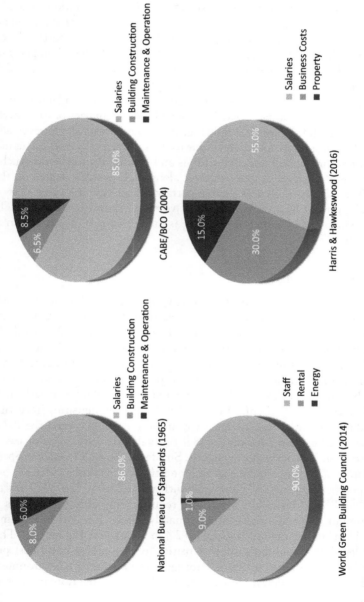

*Figure 2.3* High-level breakdown of business costs.

generate revenue in the order of three times their income. In such organisations, just a 5% increase in performance, reflected in pure revenue, could off-set the property costs. The lower 55:15 staff:property ratio proposed by Harris & Hawkeswood indicates that a performance gain of approximately 10% would be required in a corporate to offset the property costs. The corollary is that, with a focus on property costs alone, ill-advised property savings could have dire consequences for the performance and success of the organisation.

One interpretation of the staff:property cost ratio is that a large saving in property costs could be negated by a small adverse impact on staff performance (which has proportionally higher cost implications). However, additional incurred staff costs may be due to the need to replace or recruit more staff, re-train staff, change their terms and conditions, restructure such as increasing the level of management, along with possibly degrading their performance. As such, it is even important to examine the value, cost and benefit, offered by a workplace initiative rather than the cost alone. The problem, as stated earlier, is that cost is much easier to quantify than the benefits.

### Destination density

Historically, the CRE industry, along with occupiers, has been "chasing density", that is, squeezing more people into the same amount of office space in the name of space and cost efficiency. Back in 2001, Gerald Eve reported the average occupational density in UK offices to be 16.3 $m^2$ per desk. The BCO's national office density study published in 2018 (Harris et al., 2018) found an average of 9.6 $m^2$ per desk, across the whole building Net Internal Area (NIA)[4] compared with 11.8 $m^2$ per desk in the earlier 2009 study (Bedford et al., 2013). Clearly, densities have tightened significantly over 20 years (by around 40%), but even the 2.2 $m^2$ difference between 2009 and 2018 is equivalent to 19% or losing two industry standard desks per occupant. Similarly, the BCO *Guide to Specification* in 2014 suggested designing for a tighter "workplace density 8–10 $m^2$ per workspace NIA" compared with the 10 $m^2$ in the 2009 edition. Sadly, UK legislation on workspace requirements does not help as the minimum space allowed is approximately 4.6 $m^2$ per person (assuming the 11 $m^3$ minimum and a standard 2.4 m floor to ceiling height). These low legislative standards allow densities to be increased and best practice ignored, often resulting in noisier, more distracting environments, with poorer air quality, higher potential for cross-infection and possible overheating.

In contrast to the UK, other European countries fare better either because land/property is less expensive or their Works Councils have prevented over-densification. Densities vary worldwide, for example, CBRE (2015) note that the space per desk halved in Asia over a ten-year period, with the lower range in each country at approximately 4.5–8.5 m$^2$ per desk. However, they found that Australia and New Zealand remain more generous with a lower density at around 8–11 m$^2$ per desk (similar to UK). CBRE considers less than 5.6 m$^2$ per desk as a clear productivity danger zone, and with less than 9.3 m$^2$ per desk, "there are risks that not all aspects of work are fully supported, particularly knowledge-based work".

Co-working spaces are also pursuing high densities. For example, Bloomberg calculated that one of the leading co-working spaces has a global portfolio with an average occupational density of 5.1 m$^2$ per desk, assuming that all desks are occupied at one time (Sidders, 2019). The density is at its highest at their Southbank facilities in London; at just 4.1 m$^2$ per desk, it is approximately half the space suggested by the BCO, which I consider to already be too little space.

In the animal kingdom, overpopulation of habitats often leads to abnormal behaviour, disease and higher mortality (Calhoun, 1962; Herman, 1969; Morris, 1969). Consistent perceived overcrowding in the workplace leads to stress and reduced performance (Altman, 1975). A high-density workspace can also adversely affect the building infrastructure including temperature, air quality and noise along with accessibility and egress. This in turn will affect wellbeing and performance. A return to lower density office spaces is long overdue.

To meet the higher densities, desk sizes have reduced. I recall my 2 × 1 m desk at my first architectural practice, providing me with a clear 2 m between those sitting adjacent to or opposite me. The current UK industry norm is 1,400 mm wide desks, and I have worked with efficiency-zealous clients insisting that 1 m wide desks would provide sufficient space! These smaller desks result in more noise distraction, infringement of personal space and higher likelihood of cross-infection. The same client was proposing shared offices of approximately 30 m$^2$ accommodating 12 staff, equivalent to 2.5 m$^2$ per desk. I pointed out that this was too far dense and would affect performance, as well as infringing on UK legislation, and suggested that seven desks would be more appropriate. They responded by telling me that would leave a lot of wasted space between the 1 m wide desks. Desk size and density are equally important and related.

As far as I'm aware, I was the first to compare the modern workplace to chicken coups. Battery-farmed hens are mostly accommodated in high density environments with poor daylight and ventilation. In contrast, free-range hens have lots of space in which they can roam and explore with access to the outside and unlimited daylight and ventilation. Battery hens are sad looking unhealthy chickens with a short lifespan, whereas free-range hens appear to be happy, healthy, inquisitive and playful chickens that live around five times as long as a battery hen. In terms of productivity, there is undoubtedly a higher yield of eggs per square metre for battery hens, but the quality of the eggs is poor, and the demand and market value of them is low compared to free-range eggs which offer a higher return on investment.

Lower density landscaped offices are the free-range workplaces compared to those high-density open plan offices with row upon row of desks in deep-plan buildings with poor daylight and lack of natural ventilation.

An alternative way of reducing space without increasing the density is to introduce desk sharing, also termed unassigned desking or hot-desking.[5] Desk sharing is one of the elements of agile or activity-based working. Early workplace strategists observed the number of desks in use in an office at any one time and realised that if desks were unassigned and shared rather than left empty, then the desks not required could be removed, thus creating spare space. This allows the remaining desks to be spaced out and more alternative work-settings to be introduced such as focus pods and collaboration areas. This results in a lower density per desk, which benefits the occupants when they are in the office. Assuming the space accommodates the same number of workers as before (but at different times), then the space per person would be similar. Some workplace consultants refer to the space per desk as "static density" and the space per person as "dynamic density".

Another approach is to introduce desk sharing but then reduce the number of desks without adding more alternative work-settings. The workspace can then be reduced so that the same number of occupants is accommodated in less space. This approach has cost benefits, but it less attractive to the occupants as the focus is on space and cost rather than human needs and performance.

Workplace consultants conduct utilisation studies to determine the number of desks occupied over the working week; they use the data to determine the required number of desks, see Burt et al. (2010) for details. Desk utilisation rates are typically around 50%, depending on the organisation and type of work. However, this does not simply mean that a desk sharing ratio of two people per desk can be introduced as contingency is built in for busy periods, crossover of staff, observation inaccuracies, team requirements, etc. Desk sharing ratios vary across teams, but across an organisation, a target desk sharing ratio of ten people per eight desks was typical for those new to agile working. With an increase in the uptake of working from home during the Covid-19 pandemic, a more typical desk sharing ratio in the future might be ten staff per six desks on average across the organisation.

With utilisation rates as low as 50%, it may also be expected that half of the space could be saved. However, notionally, it is only the desk space that can be reduced rather than other spaces such as meeting rooms, breakout and primary circulation. Furthermore, as mentioned, the desks may be better spaced and more non-desk work-settings introduced. Often, a compromise is reached; agile working with desk sharing is used to reduce the overall space required by approximately 20%–30% whilst maintaining a comfortable level of occupational density.

When conducting research on lawyer's offices, I discovered that *The Lawyer* journal carries out an annual survey of the top 200 UK law firms. They collate information on business metrics such as number of employees, fee-earning revenue and lawyer to secretary ratios along with details of the space and property costs of offices that law firms occupy. I am particularly impressed that, in 2016, they reported the fees (revenue) per square metre as a key metric and indicator of how the space is supporting the business, a similar metric to that used in retail (Oseland, 2013a). What is so fantastic about this metric is that it is an in-house established business metric that most companies will track and report monthly.

Revenue per square metre still captures efficiency, if generating more revenue in less space, but, most importantly, the focus has shifted from cost to revenue from saving to generating money. The profit, or margin, per square metre is arguably an even better productivity metric. Revenue per square metre could be readily reported by building or by department for a more granular understanding of how office facilities support business performance.

### Quantifying the benefits

*Measuring performance and other benefits*

It is often claimed that performance is not included in a cost-benefit analysis and business case for a new workplace project because it cannot easily be measured. I find this response strange because most organisations have a handle on the performance of individual workers and the organisation as a whole. In my early days as a consultant, I would offer my property clients the option of demonstrating how a new or refurbished workplace would improve performance. They inevitably asked how I would measure performance, and in response, I would suggest using the performance metrics already in use by their organisation. Admittedly, these metrics are difficult to obtain and linking them to the impact of the workplace alone is even more difficult. Nevertheless, worker performance can be measured, and such measurements have a surprisingly long pedigree.

Measuring performance dates back around 7,000 years to the Sumerians. They not only kept records of their workforce's output (stones laid, troughs dug, fields planted, etc.) but also their costs (pay, food, lodging, etc.) and the time spent working thus allowing a form of productivity to be calculated. Of course, such work was manual, and carried out by slaves, so much easier to quantify than the range of work in the modern 21st-century office.

Empirical research into the effect of the workplace on performance dates to the early 1900s, which coincides with a growing interest in *Scientific Management*, commonly referred to as Taylorism (Taylor, 1911). In the late 1980s, professional bodies debated and listed multiple means of assessing performance. I am not entirely sure why the National Electrical Manufacturers Association (NEMA, 1989) would specifically be interested in performance, perhaps, the energy crisis, but they proposed 11 indicators, see Table 2.1. NEMA's first three indicators could be objectively measured, but the latter ones are more subjective, and all the metrics relate to individual performance rather than that of the organisation. Nevertheless, the NEMA performance metrics are laudable and comprehensive.

A few years later, the American Society of Heating, Refrigerating and Air-Conditioning Engineers (ASHRAE) held a series of workshops (Levin, 1992) to develop their own comprehensive list of performance metrics, see Table 2.2. The list builds on that of NEMA by including more quantifiable metrics and ones that impact on the organisation rather than the individuals only. Some of the metrics are likely to be already measured and monitored in-house by the organisation such as absenteeism, health costs, overtime and total cost per product. Building on NEMA and ASHRAE's

*Table 2.1* NEMA's indicators of increased performance

| No. | Indicator |
| --- | --- |
| 1 | Performing tasks more accurately |
| 2 | Performing faster without loss of accuracy |
| 3 | Capability to perform longer without tiring |
| 4 | Learning more effectively |
| 5 | Being more creative |
| 6 | Sustaining stress more effectively |
| 7 | Working together more harmoniously |
| 8 | Being more able to cope with unforeseen circumstances |
| 9 | Feeling healthier and so spending more time at work |
| 10 | Accepting more responsibility |
| 11 | Responding more positively to requests |

Oseland/CIBSE (1999).

*Table 2.2* ASHRAE's indicators of increased performance

| No. | Indicator |
| --- | --- |
| 1 | Absence from work, or workstation; unavailability on telephone |
| 2 | Health costs including sick leave, accidents and injuries |
| 3 | Observed downtime and interruptions to work |
| 4 | Controlled independent judgements of work quality, mood, etc. |
| 5 | Self-assessments of productivity |
| 6 | Component skills, task measures such as speed, slips and accuracy |
| 7 | Output from pre-existing work groups |
| 8 | Total unit cost per product or service |
| 9 | Output change in response to graded reward |
| 10 | Voluntary overtime or extra work |
| 11 | Cycle time from initiation to completion of a discrete process |
| 12 | Multiple measures at all organisational levels |
| 13 | Individual measures of performance, health and wellbeing at work |
| 14 | Time course of measures and rates of change |

Oseland/CIBSE (1999).

metrics, Oseland and Bartlett (1999) categorised performance metrics according to base activities such as reading and typing, component activities like computer use and meetings, outcomes including idea generation, responsibilities, such as meeting deadlines and minimising errors, and business objectives including profit and share value.

More recently, the World Green Building Council (Alker, 2014) proposed using several objective in-house financial metrics including absenteeism, staff turnover/retention, revenue and medical costs, along with more subjective measures such as self-reported performance and medical/physical complaints. Other in-house indicators of performance recorded by organisations are income (including sales), profitability (margin), employee attrition, staff utilisation rates, fee-earning hours, share price and outputs/projects delivered on time and within budget. After an extensive literature review, Tucker et al. (2020) at Liverpool John Moores University (LJMU) categorised productivity outputs into 6 themes and 47 sub-themes. The themes included absenteeism, staff turnover/retention, wellbeing/restoration, external performance, organisational performance and task performance, including concentration and errors.

Over the years, numerous academic research studies have reported on such performance metrics, although there is more focus on individual rather than organisational performance. Furthermore, these studies have demonstrated clear relationships between office design, including environmental conditions, and performance.

There have been many reviews collating the research. The Royal Institution of Chartered Surveyors (Thompson, 2008) reviewed 57 studies of productivity, the Commission for Architecture and the Built Environment (CABE & BCO, 2006) reviewed 111 papers and Loftness (2007) refers to Carnegie Mellon University's 297 relevant case studies. In my original review of such research (Oseland, 1999), I included 27 such studies with 35 relationships between the workplace and performance. In a later meta-analysis (Oseland and Burton, 2012), I reviewed over 200 research papers and analysed 75 credible studies with 135 quantified performance benefits. More recently, I assisted researchers at LJMU in reviewing 105 papers with 194 reported relationships that demonstrated tangible benefits of workplace design (Oseland, Tucker and Wilson, 2021).

So, there appears to be plenty of published evidence linking workplace design to performance and other benefits. And yet, it is rare that such data is used in the business case. This may be because it is difficult for businesses and organisations to find the data or more likely because, once found, they have little confidence in it. The lack of confidence is partly due to poor reliability and validity.

- *Reliability* – Refers to the consistency of the results, and unfortunately, productivity research studying the same variables can produce a range of results. In my own review (Oseland and Burton, 2012), the impact of the workplace on performance ranged from 0.3% to 68%. The range is undoubtedly due to differences in the experimental set-up and the metrics used, but it will also be influenced by extraneous factors such as motivation and reward. Herzberg (1959) distinguished between hygiene (physical) and motivational (organisational) factors. However, there are a whole host of factors that can impact on individual and organisational performance such as personal factors like experience and training and business factors such as marketing and advertising, as illustrated in Figure 2.4.

*Figure 2.4* Factors affecting performance.

- *Validity* – Relates to the accuracy of the results and how well a variable that is measured reflects what is required to be measured. Most productivity studies examine individual responses to specific performance tasks under highly controlled laboratory conditions (e.g. measuring reading performance in response to changes in desk illuminance). Such studies use metrics that are easier to measure, more directly related to design and less affected by extraneous factors. Unfortunately, these studies are most likely less relevant to the real world and operating organisations, see Figure 2.5.

Environmental psychologists refer to ecological validity, meaning how well behaviours observed under laboratory conditions reflect those in the real world, because the context of an experimental set-up is very different to the real world. More specifically, the experiments are short-term, they do not simulate all variables in the real world and an individual's motivation, attitude and so on will differ in an artificial environment, especially as many participants in such experiments are paid students. Using the zoo analogy, laboratory experiments are akin to studying animal behaviour in a caged rather than their natural environment. It is unlikely that true behaviours will be observed, and perceptions will certainly be affected by the artificial environment.

In contrast, in-house business metrics, collated in real-world environments, have high relevance to organisations but are likely to be more affected by the business and organisational factors highlighted in Figure 2.4,

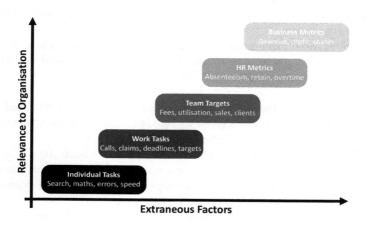

*Figure 2.5* Relevance of performance metrics.

resulting in a weaker relationship with workplace design. Nevertheless, in-house metrics are more recognised and accepted by key stakeholders, and a well-considered evaluation using control groups or benchmarks can account for the extraneous factor. Focussing on the minutiae and trying to measure individual performance rather than using good objective and existing business metrics is akin to measuring and monitoring the waggle dance of the worker bee rather than how much honey is produced by the hive.

Case studies presented at conferences, and occasionally published, tend not to provide evidence on the benefits of workplace design possibly because workplace projects are usually one-offs. In particular, the event of an organisation moving to a new or refurbished office is relatively rare and stand-alone rather than ongoing. This means that it is difficult to control for extraneous factors, such as variation in market demand or changes in personnel. This peculiar situation contrasts considerably with the retail sector which usually has multiple stores in multiple locations. I consulted a friend of mine who was a senior retail analyst at a telecommunications company, and he explained how the sales per square metre (a key performance metric) is continuously monitored across all their stores. Thus, if some intervention is made, such as a shop redesign, new branding or a regional advertising campaign, the impact on the sales per square metre can be easily monitored. If the numbers show no significant change in performance compared to other sites, then, they know the intervention failed. My friendly retail analyst considers this kind of data and continuous monitoring fundamental to running a business. In workplace, it is rare that we measure the impact of our intervention on the organisation, and when we do, they are usually a one-off measure with little context, making it extremely difficult to interpret the results and account for those extraneous factors.

In my original meta-analysis of 75 studies (Oseland and Burton, 2012), I weighted each study by three categories to determine the confidence in the performance metric in terms of direct relevance to offices:

i    *Environment* – Refers to the place where the research was carried out. A higher weighting was given to a study conducted in an actual office compared to those in laboratories or factories.

ii   *Measurement* – Relates to the metric used to measure the change in performance. A higher weighting was awarded to studies that used in-house business metrics and those employing cognitive tasks, compared to those relying on self-reported performance.

iii  *Activity time* – Refers to the amount of time that the measurement might be observed in a real office building. A higher weighting was applied to those studies measuring PC work compared to those engaged in manual labour.

After applying the weightings, the range of effect on performance was narrower, but still with up to a 4.4% increase gained for each design element. The revised, weighted, figures in these meta-analyses are more conservative and more likely to be accepted by financial directors when used in building a business case. I collaborated with LJMU to apply a similar weighting process in their meta-analysis of 105 studies conducted on behalf of the Institute of Workplace and Facilities Management, IWFM (Oseland, Tucker and Wilson, 2021).

The LJMU/IWFM weighted results for studies measuring task performance are shown in Table 2.3. There is an average range of 0.3%–3.5% increase in performance and up to an 8.2% increase for a full workplace design or refurbishment. On first appearance, the range of effects appears to be quite small, but they must be put in context of the property to staff cost ratio discussed earlier in this chapter. Increases in staff performance, such as revenue or sales, of just 5%–10% will offset the property rent and operational costs.

The LJMU/IWFM team also determined the impact of the workplace on key business metrics. Table 2.4 shows the potential impact on the broader metrics when a complete design or refurbishment is being undertaken. Caution is required when combining these broader benefits as they overlap – consider just one or two. Likewise, the potential performance increase for each of the individual design elements in Table 2.3 cannot be simply added together to calculate the overall benefit. It is more likely that the performance increase follows a law of diminishing returns, with each additional benefit diminished by one-third.

*Table 2.3* Impact of design element on task performance

| Design element | Impact on performance | | |
|---|---|---|---|
| | Lower (%) | Median (%) | Upper (%) |
| Acoustics/noise | 1.2 | 2.7 | 6.7 |
| Control | 0.6 | 1.3 | 2.6 |
| Furniture/ergonomics | 4.5 | 5.0 | 5.4 |
| IAQ/ventilation | 0.1 | 0.5 | 2.0 |
| Lighting/daylight | 0.1 | 0.5 | 1.8 |
| Space/layout | 2.7 | 3.9 | 5.2 |
| Temperature | 0.1 | 0.4 | 1.8 |
| Workplace design/refurbishment | 0.9 | 2.8 | 8.2 |
| All task performance studies | 0.3 | 1.5 | 3.5 |

*Table 2.4* Impact of workplace on performance metrics

| Metric | Impact on performance | | |
|---|---|---|---|
| | Lower (%) | Median (%) | Upper (%) |
| Absenteeism and presenteeism | 0.1 | 0.1 | 0.4 |
| Staff attrition and retention | 0.3 | 1.9 | 4.5 |
| Organisational/external performance | 2.4 | 4.0 | 5.0 |
| Wellbeing and health | 0.1 | 0.2 | 0.6 |
| All metrics | 0.1 | 0.3 | 1.6 |

*Using feedback as a performance metric*

Self-reported performance is frowned upon as an indicator of actual performance as it is not objective, thus lacking validity and reliability. On the plus side, it is a convenient indicator of performance. It is more robust if the respondent estimates their performance pre- and post-project, and the relative change is measured. Collating occupant feedback, including self-reported performance, pre- and post-project is good practice as it provides a baseline for measuring change. The relative step change in performance will be more valid than a one-off estimate.

I am still looking for a workplace study that clearly demonstrates that perceived performance is related to actual performance. However, I did find in one report of the famous Hawthorne study (Roethlisberger and Dickson, 1939) that eight wiremen of the Western Electric Company were asked to estimate their hourly output, and their actual output was also recorded. I discovered that there was a strong correlation between what they thought they produced and what they actually did, see Figure 2.6. The output of wiremen does not reflect the complexity of office work, but nonetheless, when objective measures of performance are not available, self-assessed performance can be used as an indicator, especially if the relative change is used rather than absolute value.

## Cost-benefit analysis

Cost-benefit analysis can be quite basic: a simple list of the pros and cons of each workplace option are compared against the cost, and a qualitative decision made. A more advanced analysis would include the project sponsor and project delivery team rating the various benefits and possibly weighting them by importance. The ultimate cost-benefit analysis is when the benefits are monetised and offset against the costs to determine a net cost, return on investment or payback period. This is complex, but it is possible and provides a better, more accurate evaluation of value.

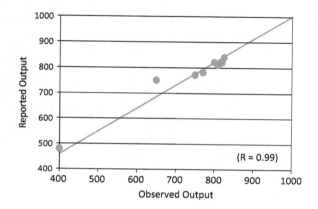

*Figure 2.6* Perceived and actual output of wiremen.

Unfortunately, as quantifying the less tangible and more qualitative benefits is considered complex, attempts of such analysis have historically been sparse. However, my own research (Oseland and Burton, 2012) recently built upon by LJMU and IWFM (Oseland, Tucker and Wilson, 2021) described in the previous section, provides a good starting point. The LJMU/IWFM data can be used to carry out a more robust cost-benefit analysis as part of the business case for a proposed workplace project, whether a new fit-out or refurbishment. This means that the business case is no longer reliant on cost alone as justification, which historically has resulted in reducing space, the infrastructure and/or facilities. This new data is by no means perfect, but it is the best available and far better than totally ignoring the impact of workplace design on performance and other benefits.

The data can be used to predict the potential performance gain expected from each design element. To monetise the benefits, the percentages could be applied to the salary of those affected or to the revenue generated by them. LJMU/IWFM suggests using their lowest percentage figure in cost-benefit analysis, representing lower confidence and a conservative prediction of the impact on performance. So, by way of illustration, when considering investing in acoustic solutions, use the 1.2% conservative figure in Table 2.3 as an estimate of potential improved worker performance to help justify any additional spend. In contrast, when faced with "value-engineering", in other words cost-cutting, use the figures to defend removing an important design element.

When building the business case for a new workplace, there are other quantifiable benefits and cost savings that can be included. For example, a move to more remote working with enabling technology (laptops, mobile devices and Wi-Fi) allows staff to work when travelling rather than wait until they return to the office. Furthermore, the ability to meet on-line means that travel will be reduced, saving on time and cost.

Oseland and Bartlett (1999) point out that many organisations tend to not link their property to performance. Some organisations, which they termed "cost-cutters", do not invest in their workplace thus potentially creating cheap ineffective working environments that degrade performance. However, even worse, are those organisations that invest heavily but perhaps in the wrong areas, thus creating costly over-priced low-performing workspaces, see Figure 2.2. Either way, it is critical that we measure and understand how the workplace affects worker performance.

## Project evaluation

Right at the start of this chapter, I mentioned that the impact of the workplace on performance is rarely assessed, resulting in lack of evidence and perpetuating the dependence on cost as the key measure of value and success. The final stage of any project should therefore be a post-occupancy evaluation (POE), and ideally, the results should be shared across the CRE industry.

While it is considered best practice to conduct a POE after completion of a workplace project, unfortunately, relatively few are carried out. POE is "the process of evaluating buildings in a systematic and rigorous manner after they have been built and occupied for some time" (Preiser, Rabinowitz and White, 1988). The emphasis is on how the building meets the requirements of the occupying organisation so relies on the feedback of the staff; however, more objective measures such as quantified performance metrics and physical monitoring may be used in the evaluation. A POE is usually conducted 6–12 months after project completion and is recommended by professional bodies such as the BCO, Royal Institute of British Architects (RIBA) and the *WELL Building Standard* (*WELL*).[6] A POE is primarily used to assess if the building is performing as intended and if value was delivered, and it also provides valuable feedback and data for use in future projects.

Many workplace projects are one-offs, but some organisations, such as those with large global property portfolios or campuses, may have regular workplace projects. This provides an opportunity to evaluate each project and measure the impact on the potential benefits. The results of the evaluation, which are specific to the organisation, can then be used in the business case and cost-benefit analysis of future projects rather than depend on generic values derived from research. Another option for a proposed large workplace project is to first conduct and evaluate a pilot project. The results from an initial smaller project will be welcomed in the business case.

The primary objective of a zoo enclosure is to benefit the wellbeing of the inhabitants and enable them to flourish rather than simply survive. Animal enclosures are not purely designed just to save space. We need to shift away from designs constrained by a space and cost focus, the disappointing *Workplace Zoo*, to ones that enable the inhabitants to flourish and thrive rather than simply survive. The CRE industry may not fully realise it, but property is a people business. The next chapter explores our innate and emerging human needs.

## Notes

1 Incidentally, the story of the Holy Grail is a "monomyth" – a folk tale that is found throughout many cultures. The moral of most monomyths is that a hero ventures forth on a quest to seek a supernatural wonder but returns the wiser, even if they did not find what they went in search of. I think we have been on the quest for the link between office design and productivity for too long, but despite the findings along the way, many of us appear none the wiser.

2 The 1%–3.5% increase in productivity was derived from my estimate of how much workplace design could potentially increase office worker performance, but unfortunately, Wheldon applied it to all UK workers regardless of their workplace or the condition of it. The £70 billion figure reported is therefore an overestimate, but as intended grabbed the media's attention.

3 Value could be expressed as a simple ratio: $V = Q \div C$, where Q refer to quality and/or quantity of a product or services and C to the cost of providing it. The value equation has some similarities to the productivity equation.

4 The NIA is the usable floor area within a building measured to the internal face of the perimeter walls on each floor; it excludes lobbies, atria, toilets, stairwells, elevators, plant rooms, columns, ducts, etc.

5 I prefer the phrase desk sharing to hot-desking as the latter was derived from the navy term "hot-bunking". Apparently, submariners work in shifts and share bunks, and a bunk would still be warm from the previous occupant.

6 The *WELL Building Standard* is a best-practice certification process, developed commercially by Delos (2015), rather than a national standard. The original certification has been replaced by *WELL v2*.

# 3 Psycho-what?

For an organisation to perform well, its people must perform well. For the workplace to enhance our wellbeing and performance, we must first understand our individual as well as group needs and how to accommodate them.

Basic human needs partly depend upon our psychology and physiology. Historic studies in early psychophysics, environmental psychology, evolutionary psychology and related disciplines clearly illustrate how the built environment affects our perception and behaviour and, in turn, our wellbeing and performance. The studies reveal group trends and also variation in individual needs and preferences due to personality, activity, context, attitude and other personal factors.

For success, workplaces need to cater for individual preferences rather than assume a homogenous design and layout will suit all. This chapter presents a summary of the psychological research evidence that I consider most relevant to workplace design.

## Early psychophysics

Gustav Theodor Fechner (1860) founded the field of psychophysics, but Fechner's work is based on the earlier research of his mentor Ernst Heinrich Weber. Psychophysics is the study and measurement of the relationship between physical stimuli and the sensory response that they produce in humans. There are a wide variety of stimuli that have been studied including temperature, light, sound, air quality, visual images, weight, space, ions, electromagnetic fields and even electric shocks. Regarding the response, early psychophysicists focused on perception such as the detection and magnitude of the stimuli.

Weber and Fechner's early experiments quantified the response of the five main senses to various stimuli. Weber found that the perceived change, or just-noticeable difference, for a particular stimulus is proportional to the initial level of that stimulus. Later, Fechner discovered that the relationship

DOI: 10.1201/9781003129974-5

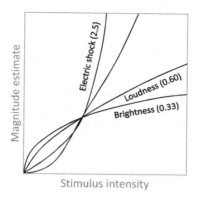

*Figure 3.1* Example of responses to stimuli (derived from Stevens, 1961).

between a stimulus and a response was non-linear, such that as the stimulus increases, the difference in response diminishes. The relationship is captured in the Weber–Fechner law:

$$S = k \log I$$

where S = $\underline{S}$ensation magnitude, I = stimulus $\underline{I}$ntensity and k = constant.

Stanley Smith Stevens (1961) conducted numerous experiments to illustrate a curvilinear relationship between the increased intensity of a physical stimulus and its estimated magnitude, known as Stevens's Power Law, see Figure 3.1.

Psychophysicists prefer to study a controlled and narrow range of variables; they focus on one element of the stimulus and a basic response to it. For instance, daylight produces light, but it has other characteristics and qualities that may illicit a different response. Daylight has a particular colour spectrum that stimulates a physiological response in humans, but a sunrise on a beautiful day can also evoke a strong emotional response. Clearly, the human response to physical stimuli is more than simple perception. The response includes a mix of psychological reactions such as sensation, comfort, emotion, mood, experience, performance and behaviour. It also includes physiological ones which may be measured through galvanic skin response (GSR), electro-encephalography (EEG) and electrocardiograph (ECG) or changes in levels of hormones or neurotransmitters.

In its broader sense, psychophysics is the study of how people perceive, interpret and react to the world. An external stimulus, such as light or sound, is first perceived through physiological and psychological processes using receptors, such as the eye or ear, but then, mental processing (cognition) allows the stimulus to be interpreted. The interpretation is affected by personal factors such as attitude and personality as well as situational ones like activity and context. As the same stimulus will provoke a different reaction in each recipient because of personal factors, it is important to measure the psychophysical response rather than just the physical stimulus. Thus, we need to measure thermal comfort rather than temperature, noise rather than sound level and perceived space rather than size, etc.

Furthermore, the senses can be easily tricked distorting perception, as demonstrated by numerous optical illusions. For example, the classic Müller-Lyer illusion illustrates how the line between the two arrow heads looks shorter than the line between the two arrow shafts even though both lines are actually the same length, see the left-hand side of Figure 3.2. Optical illusions work because the brain is making assumptions based on past experiences of regularly perceived environments. The two lines in the Müller-Lyer illusion possibly relate to the corners of rooms. Our perception is impaired because the brain struggles to reconcile the presented illusion with the expected reality. So, even at its most basic level, our response to a physical stimulus is not straightforward and is influenced by other non-physical factors.

The belief that a particular stimulus will result in a specific response reflects a rather deterministic view of behaviour. Such a view is also shared to some extent by advocates of architectural determinism who claim that the built environment has a direct and primary effect on behaviour. However, Broady (1966)[1] and other architects do not believe that architecture can directly control behaviour, but they may believe that it can influence behaviour in a predictable way. More advanced thinking is that the built environment along with personal factors can influence behaviour but not guarantee a predetermined response.

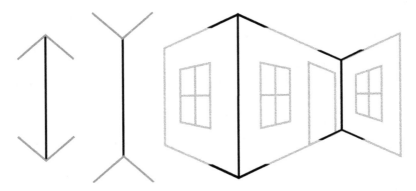

*Figure 3.2* Müller-Lyer illusion.

I have introduced psychophysics for five main reasons:

i   To highlight that the most basic human response to environmental stimuli is not linear. There comes a point when adding more of a stimulus, for example, light, doesn't make a difference and may even have a detrimental effect. Furthermore, it can't be assumed that, for example, doubling the stimulus will result in twice the response.

ii  Personal and situational factors affect the interpretation of a stimulus and the proceeding behaviours.

iii In design, we need to measure the psychophysical response of people, not just the physical stimulus. As such, measuring and making predictions from temperature alone will not guarantee thermal comfort. Likewise, measuring sound or other acoustic parameters can only partially predict a response to noise.

iv  The environment/stimulus alone does not determine behaviour but may influence it.

v   Whilst there are individual differences in responses, there are also some similarities which at least help to determine a broad acceptable range of stimuli.

## Environmental psychology

Environmental psychologists believe that the simple stimulus-response relationships observed by psychophysicists do not reflect the true behavioural response. They also place more importance on internal mental processes. They build on early psychophysics by focussing more on the personal factors and individual differences, as they recognise that the response to a physical stimulus also depends upon the people responding. Environmental psychologists are also concerned with how human behaviour and attitudes affect the environment.

The *Journal of Environmental Psychology* defines the field as "the scientific study of the transactions and interrelationships between people and their surroundings" (JEP, 2020). It is a relatively new field of psychology that became prominent in the late 1960s. However, its principles date back to the 1930s, thanks to social psychologist Kurt Lewin. Traditional psychologists took the view that behaviour is simply a deterministic response to the physical world, but Lewin proposed a different perspective, summed up by his formula:

$$B = f(P, E)$$

where B = $\underline{B}$ehaviour, P = $\underline{P}$erson and $\underline{E}$ = Environment.

Lewin highlights that human behaviour is a function of the person (for example personality, motivation and needs) as well as the physical environment. He observed that different people can perceive the same physical environment quite differently, and an individual can have different responses to the same environment at different times. It seems that our experience, our expectations and our evaluation of the current situation all affect how we interpret and respond to physical stimuli.

Environmental psychologists expanded upon Lewin's earlier proposition. Roger Barker (1968) introduced the concept of 'behavioural settings' where the pre-conceived social etiquette associated with a particular setting unconsciously influences the behaviour in that setting. For example, consider the typical behaviour of people in churches and libraries. Based on experience, we know the expected behaviour and usually conform. Barker and colleagues at the University of Kansas also pointed out that laboratory experiments and simulations, exploring and responses to the physical environment, lack ecological validity. As our response is affected by so much more than the physical stimulus, for example, experience, expectations, motivation, mood and control, the response in an artificial setting, such as a laboratory or simulation, cannot truly represent that of the real world. As such, environmental psychologists place more trust in real-world experiments – part of the reason for the weightings applied to my own research on performance discussed in Chapter 2.

Overtly studying people in the real world may also affect their responses and behaviour. The classic Hawthorne study refers to observations of various electrical assembly workers at the Western Electric Company factory carried out by Elton Mayo and colleagues (Mayo, 1933; Roethlisberger and Dickson, 1939). It was initially (possibly erroneously) reported that reducing the lighting levels had no detrimental impact on worker output because the participants were motivated by the fact that someone was taking an interest in them. A later explanation suggested that the participants were studied in a room set aside where they sat amongst a small group of their work friends. Either way, the workers were affected by extraneous factors, and the social arrangement of the space seemed as important to them as the physical parameters.

Office occupants will have different experiences and expectations of the same environment. Furthermore, the way they interpret and respond to it will be quite different, even if outwardly they appear to be behaving similarly. Designers and building managers need to recognise that what is considered a good environment by one person, including themselves, may be perceived quite differently by another. Designers are often surprised when the space is not quite used as they intended, but we all have different requirements and will either adapt our behaviour or adapt the space to make it more suitable and work better for us.

### Spaces for people

Many environmental psychologists have focussed their efforts on under-standing the effect of space – a psychophysical variable. Take optical il-lusions, such as the Müller-Lyer illusion (Figure 3.2), that illustrate how the senses can be easily tricked. They work because the brain is making assumptions based on the past experiences of regularly perceived environ-ments. In an office simulations, Edward Sadalla and Diana Oxley (1986) found that more rectangular rooms were consistently estimated as larger than less rectangular (squarer) rooms of equal size, which they termed the "rectangularity illusion". Unfortunately, Sadalla and Oxley did not inde-pendently explore the impact of ceiling height or volume and kept both constant. Anecdotal evidence indicates that offices with high ceilings ap-pear more spacious; high ceilings can also help mitigate noise and build-up of heat. For an office to be attractive, the size needs to provide a sense of space rather than simply be centred on saving space.

Osmond (1957) introduced the terms "sociofugal" and "sociopetal" space, later popularised by Somner (1967). Sociopetal space is designed to foster social interaction, and sociofugal space discourages it. A non-office example is the design of the MacDonalds' fast-food establishments. I recall the days when the seating was uncomfortable and arranged in rows accompanied by harsh, stark surfaces, decor and lighting – a space designed to grab food and go. Nowadays, the (sofa) seating is more comfortable, the lighting more subdued and the space conducive to hanging around. Osmond and Som-ner's research explored seating arrangements, but their nomenclature also applies to circulation routes, screens, environmental conditions, etc.

In his *Proxemic Framework*, Edward T. Hall (1963) calculated the pre-ferred distances required between people interacting with each other. He proposed four interpersonal distances which he termed intimate (<0.5 m), personal (~0.5–1.2 m), social (~1.2–3.7 m) and public (~3.7–7.6 m). When people infringe our personal space distance, usually reserved for family and close friends, we can feel uncomfortable. This may be due to the innate fight or flight response being triggered as the space violation is interpreted as a threat.

Interestingly, the work of Osmond and Hall was influenced by the stud-ies of zoo biologist Heini Hediger, who discovered standard interaction distances between animals, which he termed personal, social, critical and flight. Hediger proposed that the quality of an animal enclosure is as impor-tant as the quantity (space). If the enclosure represents the natural environ-ment of its inhabitants, then their biological and ethological requirements will be met, such that the animals feel safe, satisfied and flourish.

Our real-life experience reflects Hall's hypothesis: think about where people choose to sit on park benches, on public transport or in cafés; the preferred seat is quite predictable. However, I was once told by an overseas colleague that the "normal" rules do not apply overseas and those from Eastern cultures prefer to sit at crowded tables rather than alone, as imitated by Wagamama-style seating. Indeed, relatively recent research has confirmed that interpersonal distances vary by culture (Sorokowska et al., 2017).

On a related topic, Steelcase (2012) demonstrated that some cultures are more tolerant of high-density working environments than others. Steelcase plotted office space requirements for several countries using six cultural dimensions and found significant differences – in particular that some countries are more collaborative at work than others and some require higher levels of privacy and confidentiality. Although many organisations are now global and may even carry out similar work across countries, it does not mean that the same standardised workspace will suit employees from different cultures, countries and traditions. Designers and workplace strategists need to respect and accommodate cultural differences when considering layouts and density.

Personality traits (such as introversion/extroversion) and other personal factors (such as gender and age) will all affect the preferred interpersonal distance. Situational factors are also important – an assessment of the situation affects the acceptable personal space. Consider that crowding on public transport is uncomfortable but tolerated, whereas a crowded sports stadium or music venue may be more welcomed as it adds to the atmosphere. Incidentally, the Covid-19 pandemic and risk of cross-infection is also likely to have affected views on personal space. In my own research (Oseland and Donald, 1993), I found that perceived space and privacy was dependent on where you are (place/location), who you are with (people/company) and what you are doing (process/activity). The perceived space and privacy were also affected by psychological and personal factors and, of course, the physical environment, see Figure 3.3. Admittedly, my research

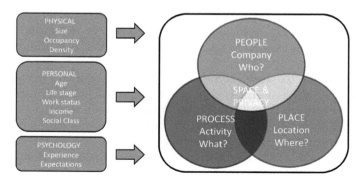

*Figure 3.3* Factors affecting perceived space and privacy.

was conducted in homes, but the results are in-line with broader environmental psychology theories and so transferable to other spaces, like offices, and to other physical stimuli.

Environmental psychologist Dan Stokols distinguishes between high density and crowding. Density is the measured number of people per area, whereas crowding is perceived and subjective; it is a psychophysical response dependent on the amount of surrounding people but also related to personal, cultural and situational factors such as the activity being conducted (Stokols, 1972). Like personal space, a high-density environment may not feel crowded for certain activities, like a trading floor, but the same density may feel crowded for other activities. Rooms with more daylight and lighter colours are perceived as less crowded, whereas tasks requiring interaction are degraded in high density environments. One coping strategy deployed when space is infringed is for people to mark out their territory using symbolic and physical boundaries such as cabinets, documents and personal artefacts such as family photos and mascots.

Irwin Altman unified various theories of space, including personal space and crowding, together into his "privacy regulation theory" (Altman, 1975), see Figure 3.4. Altman conceptualised privacy to be a dialectic and dynamic process for controlling the level of availability to others rather than the usual simple view that privacy is a state of social withdrawal. "Dialectic" refers to whether people are actively seeking or avoiding social interaction, and "dynamics" means that the desired level of interaction varies according to individual differences, situations and circumstances over time. According

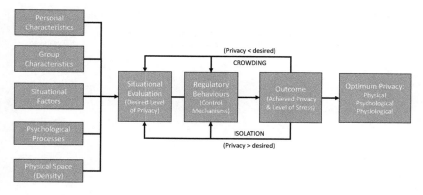

*Figure 3.4* Privacy regulation theory (adapted from Altman, 1975).

to Altman, not achieving the desired level of privacy will result in discomfort and stress; too little privacy produces feelings of overcrowding, whereas too much privacy results in social isolation. Furthermore, he suggests that people use control mechanisms or coping strategies to control their level of privacy such as marking territoriality leaving personal items out or building barriers out of books, plants or filing cabinets.

## Evolutionary psychology

A relatively new field of psychology is evolutionary psychology, dating back to the mid-1970s but not widely recognised until the 1990s. The theory of evolution acknowledges that over time, animal physiology develops, through natural selection, to ensure the survival of the species. The study of human evolution tends to focus on advances in our physiology and intelligence. The biological advancement of a species is a slow process taking millennia, but over time, our physiology has evolved and adapted for survival and wellbeing.

Evolutionary psychologists assert that our innate human behaviour and psychological processes have also evolved over time to aid survival. For around 300,000 years,[2] Homo sapiens evolved to live and survive on the African plains, termed the Savannah Hypothesis (Orians and Heerwagen, 1992), whereas modern humans have only worked and "survived" in unnatural large office buildings for 100 years or so, see Figure 3.5. Consequently, psychological development and adaptation to this relatively new environment are out of sync. In other words, modern offices do not meet our evolved innate human needs, in turn, affecting our comfort, wellbeing and performance.[3]

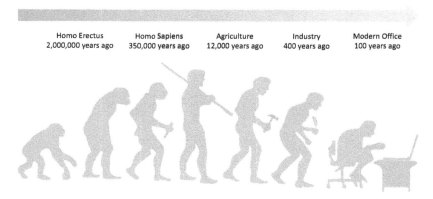

| Homo Erectus | Homo Sapiens | Agriculture | Industry | Modern Office |
|---|---|---|---|---|
| 2,000,000 years ago | 350,000 years ago | 12,000 years ago | 400 years ago | 100 years ago |

*Figure 3.5* Evolution and working environment (adapted from Kissclipart).

*Figure 3.6* Humans evolved to survive on the African Savannah (Byrdyak, CC BY-SA 4.0, via Wikimedia Commons).

The field of evolutionary psychology is not without its critics. People often misinterpret Darwin's theory of evolution as saying it is the strongest of the species that survives, but a more accurate interpretation is "the species that survives is the one that is able best to adapt and adjust to the changing environment in which it finds itself" (Megginson, 1963). While we have innate preferences, we also adapt our environment to suit us or adapt our behaviour to it when those preferences are not met. As my old colleague Nick Pell succinctly puts it, "we humans are adaptable creatures of habit, inquisitive by nature and intrinsically wired to evolve" (Pell, 2021). Second, the Pleistocene or Ice Age lasted from about 2.5 million to 12,000 years ago, whereas Homo sapiens may have moved north around 150,000 years ago and so had to adapt to the harsher/colder environment. Therefore, the Savannah (Figure 3.6) is not the only environment that influences our preferences.

Nevertheless, there is a significant body of research, particularly in the fields of neuroscience and biophilia, to support the Savannah Hypothesis. The design implications of biophilia, or "biophilic design", are covered later, but an introduction to the field is first required as is an introduction to neuroscience.

### Biophilia

A key theme within evolutionary psychology is biophilia, which succinctly put is our innate affinity to nature. The term was popularised by biologist

Edward O. Wilson but was coined ten years earlier by psychoanalyst Erich Fromm. In his book *The Anatomy of Human Destructiveness* (1973), Fromm described biophilia as "the passionate love of life and of all that is alive". Some ten years later, Wilson's book titled *Biophilia* (1984) provides an intensely personal account of his response to nature, and he defines biophilia as "the urge to affiliate with other forms of life". He passionately believes that our natural affinity for life is the very essence of our humanity and binds us to all other living species.

Stephen Kellert, in the introduction to the seminal book *Biophilic Design: The Theory, Science and Practice of Bringing Buildings to Life* (Kellert, Heerwagen and Mador, 2008), reminds us that "biophilia is the inherent human inclination to affiliate with natural systems and processes, especially life and life-like features of the nonhuman environment". However, Kellert continues by explaining that our affiliation to nature "became biologically encoded because it proved instrumental in enhancing human physical, emotional, and intellectual fitness during the long course of human evolution" and as such "People's dependence on contact with nature reflects the reality of having evolved in a largely natural, not artificial or constructed, world". This doesn't mean that we don't desire more readily available shelter, food, comfort and Wi-Fi or that we are not highly adaptive creatures, but, perhaps, we are to some extent "hard-wired" for a more natural environment. We not only like nature but also need it to flourish and perform well.

Evolutionary psychologists Rachel and Stephen Kaplan have promoted the benefits of biophilia since the early 1990s with their introduction of Attention Restoration Theory (ART). They propose that mental fatigue is reduced whereas our ability to focus and concentrate is improved with exposure to natural environments. The capacity of the brain to focus on a specific stimulus or task is limited and results in fatigue of "directed attention". However, when exposed to nature, people feel refreshed and replenished because nature provides a setting for "non-taxing involuntary attention" enabling our directed attention capacities to recover.

Many research studies support the restorative powers of nature. For example, Ohly et al. (2016) conducted a meta-analysis of 31 studies and discovered that scores on several attention tasks increased significantly after exposure to natural settings. Heschong et al. (2004) conducted a study of a call centre, with good embedded objective performance metrics, and found that operatives with views out, access to daylight and greenery processed calls 6% to 12% faster than their colleagues in poorer spaces. Researchers at Kansas University (Atchley, Strayer and Atchley, 2012) found that performance on lateral thinking, or creativity, tests increased by 50% after the participants spent four to six days back-packing in the American wilderness. Simone Ritter (2012) found that lateral thinking test scores increased by 15% when the participants were exposed to new experiences, such as different foods, an alternative route to work and meeting new people. She suggests that people exposed to a standard routine become stale and suffer from "functional fixedness", whereas new experiences unlock the synapses in the brain and create new neural pathways.

Physiological benefits of biophilia have also been found. For example, a study of Shinrin-Yoku, Japanese for "forest bathing", showed that participants immersed in nature, compared to those in the city, had a reduced pulse rate, reduced blood pressure and reduced cortisol levels plus increased immune function – all indicators of lower stress levels (Park et al., 2010).

Similar biophilic benefits have been found indoors. A fellow psychologist Craig Knight and his colleagues have run a series of experiments where they found that environments with plants (and art) result in increased creative ideas compared with "unenriched" workspaces (Knight and Haslam, 2010). Bill Browning, an environmental strategist, has documented the benefits of biophilia for some time and provides clear evidence for the benefits in his books *The Economics of Biophilia* (2012) and *14 Patterns of Biophilic Design* (2014). The latter book also includes clear guidance on biophilic design.

Biophilic design, that is designing workplaces that introduce nature or better connect the occupants to nature, has become a popular with architects and the workplace design community since around 2015. Biophilic design captures a broad range of elements in-line with evolutionary psychology principles. However, more recently, biophilic design has been reduced, by some companies, to merely providing landscaping and plants in the office. Such organisations proclaim that potted plants plonked on a desk will reduce air pollutants and are the panacea for enhancing wellbeing and improving performance. I recommend plants in the office, but my issue with these bold claims is two-fold. First, there is so much more to biophilic design than plants, and second, there are much more efficient ways of reducing air pollutants than adding plants alone.

I need to mention biomimicry at this point because there is some overlap between biomimicry and biophilic design. The Biomimicry Institute states that "biomimicry is a practice that learns from and mimics the strategies found in nature to solve human design challenges". It tends to refer to engineering rather than architecture and favours structures and processes found in nature that are also sustainable as they use natural resources and energy. Classic examples of biomimicry include the nose of the Shinkansen (bullet train) shaped like a kingfisher's beak, boat hulls covered with material based on shark skin and passive stack ventilation systems based on termite mounds. Nevertheless, biomimicry is sometimes used to refer to shapes and patterns found in nature and copied in design, such as hexagons from beehives, helixes from seashells and fractals from Romanesco broccoli or snowflakes.

### Neuroscience

Neuroscience is the scientific study of the structure and function of the human brain and nervous system, especially, neurons and neural circuits. Neuroscientists explore psychological processes such as perception, memory and behaviour but take a biological and physiological perspective. Only in recent years have neuroscientists turned their attention to behaviour in the workplace, and there are strong links between neuroscience and evolutionary psychology.

Basically, part of the brain, termed the limbic system, will unconsciously respond to environments that are threatening, resulting in discomfort and stress. It therefore follows that environments which do not meet our innate basic needs will have a similar underlying and ongoing, albeit slightly less intense, effect.

The "Triune Brain" is a simplified model of the development of the brain proposed by neuroscientist Paul D MacLean in the late 1960s (but not fully documented until his 1990 book). The model may have some bearing on Evolutionary Psychology and our innate response to the surrounding environment. MacLean proposes that the triune brain consists of three distinct structures that developed over time and were sequentially "added" to the brain as part of our evolution. The three structures in order of development are (i) the reptilian complex, (ii) the paleomammalian complex and (iii) the neomammalian complex, see Figure 3.7.

The reptilian complex, also known as the primitive or survival brain, is the oldest part of the brain, based in the brain stem and cerebellum, and

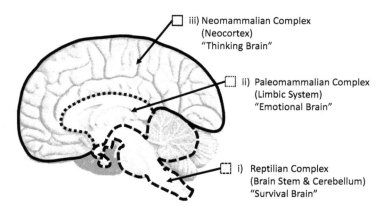

*Figure* 3.7 The Triune Brain system (adapted from Looie496, public domain, via Wikimedia Commons).

it controls basic unconscious (autonomic) functions including breathing, heart rate, digestion, pupillary response and urination. The paleomammalian complex, or emotional brain developed in the first mammals, lies in the limbic system (including the amygdala, hypothalamus and hippocampus) and is responsible for our "fight or flight" response. The amygdala makes a superfast unconscious evaluation of a situation and initiates a rapid response to avoid danger and threat to survival. The neomammalian complex, neocortex or thinking/rational brain supposedly first developed in primates and culminated in humans. It is the part of the brain that offers all high-order functioning such as cognition, learning, memory, logic, imagination, problem-solving and language.

However, the limbic system is not 100% accurate, and negligible threats occasionally result in an excessive and disproportionate response. As such, when under duress at work, receiving advice from a colleague might be perceived as overt criticism and spark an unwarranted overreaction – "the straw that broke the camel's back". This is sometimes referred to as "going limbic", "amygdala hijack" or "releasing the inner chimp". It is therefore feasible that the limbic system may unconsciously respond to environments that do not meet our basic needs or are considered a threat, perhaps resulting in discomfort, fatigue, stress, poor wellbeing and ultimately loss of performance.

The *Triune Brain* model has received criticism from modern neuroscientists. The evidence to demonstrate that three distinct parts of the brain developed sequentially over time is lacking. Nevertheless, the model offers a basic, easy to digest, insight into brain functioning and our response to different types of stimuli.

### Anthropology and Dunbar's number

There is some overlap in the interests of psychologists and anthropologists. Whilst classic psychology is more concerned with studying individuals, anthropology explores humans within their community, society and culture. Psychologists focus on internal processes such as perception, memory and behaviour, whereas anthropologists are concerned with broader issues like the origin of humans, development of language, health and nutrition. So, the field of anthropology is relevant to designing for innate human needs.

For example, anthropologist Robin Dunbar proposed the "rule of 150", referred to as Dunbar's Number. This is about the size of a human social network and the limit to the number of people that we can recognise and have a meaningful social relationship with, due to our cognitive limit. Dunbar derived his number by studying primates and their social group sizes. He correlated the neocortex size of a range of primates with their social group size, then extrapolated that for humans and found it to be 147.8 and generalised to 150 ± 50 (Dunbar, 1992). The number has been corroborated by the size of basic hunter–gatherer communities, the average village size in the *Domesday Book*, Christmas card lists and military units (the company or Roman maniple).

Dunbar's Number impacts on the size of teams, organisations and floor plates in office buildings. An unwritten management rule is that organisations below 150 do not need a highly hierarchical management structure, and apparently that is partly why Gore, the manufacturer of Gore-Tex fabric, only builds self-contained factories that contain up to 150 workers.[4] Dunbar believes that, in organisations above 150, the management and workforce will not all know each other, resulting in a loss of flow of information and a reluctance to help each other out. Many offices today are built with huge floor plates containing 400–600 desks. This is beyond the human scale for meaningful relationships and is likely to impact on management, flow of information and performance.

Dunbar introduced other bands of numbers reflecting levels of acquaintanceships: 5 for clique or close friends, 15 for sympathy or extended family, 50 for clan or close network, 150 for friendship or personal network, 500 for tribe and 1,500 for community. Note that each level increases approximately by a factor of three. The number recommended for the optimal team size varies. Those in technology who have adopted their version of agile working will be aware that the recommended optimal size is 7 ± 2, whereas other researchers have suggested from 5 to 12. A contubernium, the smallest unit of the Roman army, was composed of eight legionaries with a mule, and in the modern army, a squad is composed of 8–12 soldiers. Therefore in organisations where team members work closely together, clusters or zones of 8–12 desks will be more useful.

## Motivation theory

### Maslow

A branch of psychology worthy of highlighting separately is motivation theory. *Maslow's Hierarchy of Needs* (1943) is a metatheory familiar to all budding psychologists. Maslow proposed that, for humans to perform to their maximum capability, several levels of needs must be met in ascending order; see the ladder diagram in Figure 3.8.[5] The lower-order needs refer to basic physiological needs such as comfort and nourishment along with safety needs including financial stability as well as protection and security, all required for human survival. If these fundamental needs are not met, then our performance is inhibited. Providing comfort, nourishment, protection

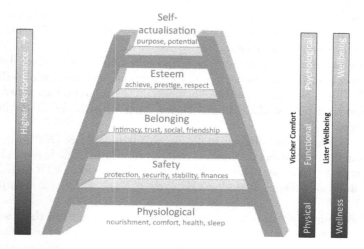

*Figure 3.8* Maslow's hierarchy of needs.

and safety are also the basics of zoo enclosure design. Much of the provision is generic, but it should also be tailored for specific individual needs. In contrast, the higher-order needs refer to more psychological, emotional and social factors. Interestingly, another core provision for animal enclosures is the opportunity to express normal behaviour – "accommodation should take account of the natural habitat of the species and seek to meet the physiological and psychological needs of the animal" (DETR, 2000). To reach our maximum potential, we need a sense of belonging, a sense of achievement and a sense of purpose. Maslow may have been referring to broader societal needs and maximising potential, but they also transfer to the workplace (both organisation and space) and maximising performance.

Jaqueline Vischer (2004, 2005) distinguished between physical, functional and psychological comfort in her *Habitability Pyramid*, with occupant satisfaction and wellbeing at the top of the pyramid. Physical comfort refers to our basic needs, which should be provided through elementary architectural design and building services, including consideration of indoor air quality, temperature, light and acoustics. It also includes access to and from the building's floors, toilets and restaurant facilities. Not meeting these needs impacts on comfort and performance. Functional comfort refers to how effective the workplace is in allowing the occupants to perform their daily work tasks. This level includes psychophysical factors such as thermal comfort and noise plus the provision of spaces supporting collaboration, creativity and focussed work. Psychological comfort acknowledges the importance of higher-order needs like environmental control, privacy and territoriality, which also link to connectedness and a sense of belonging. There is clearly some overlap between Vischer's *Habitability Pyramid* and *Maslow's Hierarchy of Needs*.

Approximately 10 years ago, the focus of the CRE industry shifted from the more pragmatic "productivity" to the softer "wellness" and "wellbeing". This is evident by the development of evaluation and accreditation methods such as WELL and *FitWell* and later with the BCO's *Wellness Matters*.

Wellbeing is quite difficult to define; as Pollard and Lee (2003) say, it is "a complex, multi-faceted construct that has continued to elude researchers' attempts to define and measure it". However, I quite like DEFRA's (2007) definition as it alludes to *Maslow's Hierarchy of Needs* and *a sense of purpose:*

> A positive physical, social and mental state; it is not just the absence of pain, discomfort and incapacity. It requires that *basic needs are met,* that individuals have *a sense of purpose,* that they feel able to achieve important personal goals and participate in society (my emphasis).

There is also some confusion in texts over the difference between wellness and wellbeing. Kate Lister (2014) suggests that wellness is focused on physical health whereas wellbeing includes wellness, but also the psychological state of the individual worker. Lister also proposes that wellness relates to the lower levels of *Maslow's Hierarchy of Needs*, whereas wellbeing overlays with the upper ones.

### Herzberg

Another often mentioned motivational theory is the *Two-Factor Theory*, also termed *Herzberg's Motivation-Hygiene Theory* (1959), which distinguishes between hygiene factors and motivators. Hygiene factors refer to status, job security, salary and working conditions (perhaps including design features), with some similarity to Maslow's middle rungs. Motivators such as recognition, responsibility, having purpose relate more to Maslow's upper rungs. A major difference between Herzberg and Maslow is that *Two-Factor Theory* suggests that if the hygiene factors are not met, then performance will decrease, whereas, to increase performance, the motivators must be met.

A major criticism of *Maslow's Hierarchy of Needs* is that it has never been empirically proven so remains a theory. Ken Raisbeck (2003), who 20 years ago was a junior colleague of mine and a promising workplace consultant, attempted to demonstrate *Maslow's Hierarchy of Needs* as part of his MBA by conducting a staff survey in the shared services department at a refurbished bank. As I was more familiar with surveys and statistical analysis, I

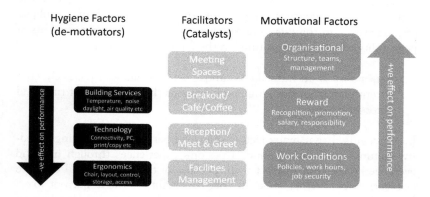

*Figure 3.9* Three-Factor Model for the workplace based on Herzberg and Raisbeck.

assisted Ken with designing the survey and I carried out a factor analysis of the data. Our analysis did not support Maslow but did have some similarity to Herzberg's *Two-Factor Theory*. However, a significant difference was that our analysis revealed three factors, which we termed hygiene, motivators and facilitators, see Figure 3.9. The motivators are similar to those found by Herzberg, whereas the hygiene factors are more related to environmental conditions, design and technology. Our third factor, the facilitators, appears to relate to the physical components of the office but ones that help motivate the staff, such as meeting and breakout spaces.

So, it appears that our innate human needs operate at different levels. The most basic (hygiene factor) level is relevant to all species and includes nourishment, safety, sanitation and the appropriate environmental conditions (temperature, air quality, daylight, acoustics, etc.) for comfort. If these basic needs are not met, then the animal will not thrive, will not perform well and may struggle to even survive. Most basic human needs have not changed since our days on the African Savannah, and yet, studies show that they are not provided for in the modern office. For humans, there are also more psychological higher-order needs to be met, and many relate to our ancestry and historical needs, such as a sense of belonging and friendship. Between the more physical and psychological needs are the psychophysical such as perceived privacy, territoriality, noise, space and control.

## Personality theory

Personality is derived from *persona* which is Latin for "mask"; this etymology implies that personality is the mask we present to the world. There is no consensus amongst psychologists on a single all-encompassing definition of personality (John, Robins and Pervin, 2008); this is partly due to the many different approaches to personality theory and breadth of the subject area. However, in the literature, there are several reoccurring elements of personality which I have captured in the following definition:

> Personality is an individual's unique set of traits and relatively consistent pattern of thinking and behaviour that persists over time and across situations.

It is recognised that many factors influence personality including heredity, culture, family background, a person's experiences through life, and even the people they interact with. Because of these factors and the core elements of personality, it seems that personality is stable but not fixed. Personality is a bias towards particular traits (characteristics) that in turn affect behaviour. This embedded proclivity for behaving in a particular way means that it is also likely that people prefer and seek out environments that support their natural mode of behaviour and underlying personality.

It is important to recognise that the modern workplace should accommodate the range of different personality types and provide the environment that enhances their wellbeing and performance, something often forgotten in the *Workplace Zoo*. Fortunately, different job functions attract similar personality types (Schaubhut and Thompson, 2012). Apparently, extroverts are better suited to sales and marketing, whereas introverts are more likely to prefer research and analysis. Therefore, the workplace for a department or business unit could be designed to suit the majority of personality types, with alternative options for less common types.

### Temperaments and psychoanalysis

Personality theories possibly date back to ancient Egypt or Mesopotamia, but the Greek physician Hippocrates (circa 400 BC) is recognised as developing the first structured theory of personality. He proposed that different personality types are caused by the (in)balance of bodily fluids, termed the four humours, see Figure 3.10. Galen (circa 150 AD) refined Hippocrates' four humours as the four temperaments. Together, they believed that phlegmatic (or calm) people have a higher concentration of phlegm; sanguine (or optimistic) people have more blood; melancholic (or depressed) people have high levels of black bile; and irritable people have high levels of yellow bile. Interestingly, these Greek temperaments are still sometimes used today to describe personality characteristics such as the super-trait theory of Eysenck and Eysenck (1975) and modern-day neuropsychologists acknowledge that the presence of certain chemicals in the brain affect mood and behaviour.

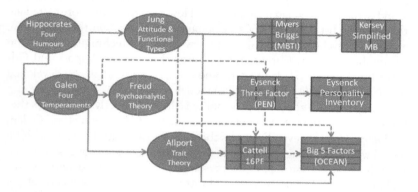

*Figure 3.10* The development of modern personality theories.

So, the notion of bodily fluids affecting behaviour is not as bizarre as it initially sounds.

The next significant stage on the development of personality theories was at the turn of the 20th century when Sigmund Freud developed the psychodynamic theory of personality. Freud highlighted the influence of unconscious factors, our past experiences and our libido or sexual drive, on behaviour. However, Freud's theories have been criticised as pseudo-scientific and sexist, and there are mixed views amongst psychologists of the current relevance of his work. One of Freud's critics was the analytical theorist Carl Jung, who developed his own psychodynamic viewpoint. Jung emphasised the future and the unconscious even more so than Freud but without a strong emphasis on sexuality. He also developed the idea of the collective unconscious, which is the belief that all people have the same basic patterns of behaviour. Of more significance was that Jung grouped people into two broad types based on their general attitude, namely introverts and the extroverts. Jung considers an attitude to be a person's predisposition to behave in a particular way. Categorising personality on an extroversion scale has influenced most subsequent theories of personality and is still very much referred to in organisational psychology and business management theory.

### Trait theory

As an alternative to psychodynamics, another approach to understanding personality is to identify and describe it in terms of traits or characteristics. The problem with trait theory is there are so many descriptors of personality that it is difficult to make sense of them. Allport and Odbert (1936) conducted a lexical approach to the dimensions of personality and initially found some 17,953 related descriptors. However, they went on to reduce this gigantic list down to 4,504 personality traits. Around the same time, a group of psychologists began using new statistical techniques to develop personality theories. Cattell (1947) managed to reduce Allport and Odbert's list down to 171 descriptors, and using factor analysis, he developed his *Sixteen Personality Factors* (16PF) model.

Eysenck's (1947) theory of personality is derived directly from Jung's theories and even relates to the four temperaments of Hippocrates and Galen, see Figure 3.11. However, Eysenck's theory is also a rebuttal to Cattels' 16PF model which he thought had too many superfluous dimensions, so Eysenck initially proposed using just two personality dimensions: extroversion (E) and neuroticism (N). Extroverts and introverts sit on opposite ends of the extroversion dimension: an "extrovert is a friendly person who seeks company, desires excitement, takes risks and acts on impulse, whereas the introvert is a quiet, reflective person who prefers his or her own company and does not enjoy large social events" (Eysenck and Eysenck, 1975). The neuroticism dimension is a dimension of emotional stability that ranges from calm and collected people to ones that experience negative emotional states such as anxiety, apprehensiveness and nervousness.

Regardless of whether people are introverted or extroverted, they need to cope with the world and will have a preferred way of doing this. Jung suggested that there are four basic ways of coping, termed functions, and when combined with one of his two attitudes, they form eight different personality types. The sensing (S) and intuition (N) function relates to the way individuals perceive and acquire information. Sensing individuals carefully examine information and employ all of their senses in their investigations;

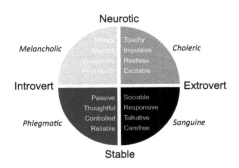

*Figure 3.11* Eysenck's theory of personality (adapted from Eysenck and Eysenck, 1975).

whereas those who are intuitive rely more on their instincts and gut-feeling. The two functions related to reaching decisions are thinking (T) and feeling (F). Thinkers are objective, analytic and logical and consider facts in reaching conclusions; in contrast, feeling individuals are subjective and consider how their decisions will impact others.

Briggs and Myers (1987) elaborated on Jung's personality theory by adding a function which indicates the way people interact with the environment. Judgers (J) prefer an organised, stable environment and strive to regulate their lives, whereas perceivers (P) are flexible and spontaneous preferring to stay open to new opportunities. Adding these dimensions to those of Jung creates a matrix of functions resulting in 16 personality types. These 16 personality types are usually referred to by the dimension acronym and common descriptors (stereotypes) for the types, see Figure 3.12.

The *Myers-Briggs Type Indicator* (MBTI), and its corresponding personality types, is popular in modern business management. The MBTI is used for evaluating and developing teams for improving communication between different personality types and to help determine the suitability of potential employees to job roles. The MBTI is viewed by some as having too many distinct personality types. Consequently, *The Big Five Personality Inventory*, or *Five Factor Model*, has gained increasing popularity due to the

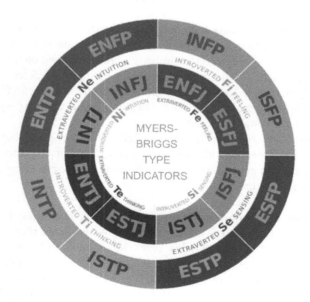

*Figure 3.12* Myers–Briggs personality-type indicator (Offnfopt, public domain, via Wikimedia Commons).

manageable number of personality traits, the practicality of its relatively short questionnaire, the robustness of the approach across time and cultures and that the five factors were determined by a number of psychologists conducting research independently and in parallel. The initial model was originally developed in 1961 but did not gain popularity until the 1980s, mostly due to the work of Costa and McCrae (1992).

*The Big Five* has its origins in trait theory verified by statistical analysis. The researchers all began by studying known personality traits and then used factor analysis on hundreds of measures of these traits in order to find the underlying five factors of personality. The five factors are Openness, Conscientiousness, Extroversion, Agreeableness and Neuroticism, often referred to as OCEAN. Interestingly, OCEAN includes Eysenck's dimensions of extroversion and neuroticism.

- *Openness (to experience)* – Reflects the range of interests and fascination with novelty; open people are creative, curious and artistically sensitive, whereas those who are less open are more conventional and like the familiar.
- *Conscientiousness* – Reflects a measure of reliability; a highly conscientious person is responsible, organised, dependable and persistent, whereas unconscientious or undirected people are easily distracted and unreliable.
- *Extroversion* – Reflects the comfort level with relationships; extroverts tend to be gregarious, assertive and sociable in nature, whereas introverts tend to be reserved, reflective and quiet, preferring their own company.
- *Agreeableness* – Reflects an individual's tendency to defer to others; highly agreeable people are cooperative, affectionate and trusting, whereas those more disagreeable are antagonistic, challenging and less empathetic towards colleagues.
- *Neuroticism (emotional stability)* – Reflects a person's ability to bear up to stress; people with positive emotional stability tend to be calm, self-confident and secure, whereas the more neurotic are nervous, anxious and insecure.

My literature review of research into team performance and collaboration (Oseland, 2012) revealed that the most effective teams are those that consist of a healthy mix of personality types. Much research has been carried out comparing the performance of teams where the members have either similar personality profiles (homogenous group) or quite different personality profiles (heterogeneous group). Research, and arousal theory (which is discussed later), indicates that different personality types are better at different tasks, for example, introverts are more successful at routine and detailed tasks than extroverts. Table 3.1 provides a description of the five factors along with the potential implications for communication style and collaboration spaces.

*Table 3.1* Big five implications for performance and collaboration

| *Implication for task performance* | *Implication for collaboration* |
| --- | --- |
| **Openness vs Conservative**<br>Evidence supports the importance of openness for creative and imaginative tasks but suggests that openness is less important, or even detrimental, when the task is of a more routine nature. | **Open** people prefer face-to-face meetings, brainstorming, plus stimulating, different and new spaces.<br>**Conservative** people prefer formal, familiar, conforming and traditional spaces. |
| **Conscientiousness vs Casual**<br>Should be positively related to team performance across a wide variety of tasks and settings, | **Conscientious** people prefer planned, formal, well-organised, minuted meetings.<br>**Casual** people prefer impromptu and informal meetings, idea generation and quick interactions. |
| **Extroversion vs Introversion**<br>Extroversion is related to team performance when tasks involve imaginative or creative activity but may inhibit performance when tasks call for precise, sequential and logical behaviour. | **Extroverts** prefer face-to-face and socialising, large social groups plus impromptu, informal, off-site meetings and stimulating spaces.<br>**Introverts** prefer written communications, distributed information, small groups, teleconferences and subdued spaces. |
| **Agreeableness vs Challenging**<br>Agreeableness may be important for performance in long-term teams with tasks that involve persuasion or other socially related dimensions; when tasks do not require a high degree of social interaction, agreeableness may inhibit performance. | **Agreeable** people prefer large meetings with structure and distributed information to help gain group consensus.<br>**Challenging** prefer unstructured face-to-face meetings where they can challenge and derail. |
| **Neuroticism vs Emotional Stability**<br>The level of emotional stability in the team correlates with team performance for a wide range of tasks. | **Neurotic** people prefer well-planned formal meetings with advance notice and information; also subdued environments.<br>**Stable** people are comfortable with large, impromptu or informal meetings. |

Heterogeneous groups convey a more varied style of problem-solving and interact more. Furthermore, they discuss alternative solutions, devise more creative ideas and are found overall to be more effective. In contrast, high cohesiveness through homogeneity can lead to "groupthink". This is when like-minded team members shut themselves off from outsiders with conflicting views and develop an unrealistic sense of righteousness and blinkered views and solutions. In contrast, heterogeneous groups challenge each other but are more likely to develop a more unique, effective and creative solution. Whilst heterogeneous teams take longer to bond, they ultimately deliver more creative, innovative, well-considered and successful outputs (Briggs, Copeland and Haynes, 2007).

Bill Gates, Gandhi, Einstein, Lincoln, J.K. Rowling, Darwin, Virginia Woolf and Chopin have all made significant contributions to society through the arts, science, business or politics. Another thing they have in common is that they are all classic introverts. Susan Cain, in her book *Quiet: The Power of Introverts in a World That Can't Stop Talking* (2012), reminded us that introverts are often overlooked despite making considerable contributions to society. She also reminded us that the workplace contains similar proportions of introverts and extroverts, but the modern workplace is often designed with the extrovert in mind. There is a tendency to create open plan, noisy, buzzy, crowded environments that are stimulating and facilitate interaction and collaboration. Yet, these environments can cause distraction, higher stress and poor performance for introverts, especially those involved in complex and detailed analysis, those roles that are often attractive to introverts.

Numerous psychology studies have shown that extroverts prefer stimulating environments, whereas introverts are more productive in calmer and subtle workspaces. One of my studies (Oseland, 2012) demonstrated that introverts spend more time in solitary activity, predominantly communicate using email and when they meet prefer enclosed offices and meeting rooms. On the other hand, extroverts spend more time in face-to-face communication and prefer meeting in bars, informal/social spaces and huddle rooms. I also found that those scoring high on the neuroticism scale have overlapping workplace requirements to introverts.

One explanation for the difference in requirements between introverts and extroverts is the innate level of arousal, which relates to excitement or interest. A key fundamental theory in psychology is the Yerkes-Dodson Law (Yerkes and Dodson, 1908) which proposes an inverted U-shape relationship between a person's performance and their level of arousal, see Figure 3.13. People can perform better if they are stimulated or motivated, which increases their level of arousal, but there is a limit because too much stimulation can lead to stress and thus reduce performance. In contrast, under-stimulation leads to lethargy and even fatigue thus reducing performance. Furthermore, individuals have a different base level of arousal

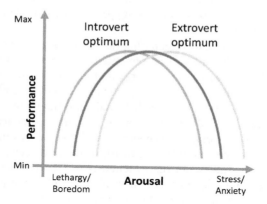

*Figure 3.13* Arousal of introverts and extroverts (adapted from Yerkes and Dodson, 1908).

and therefore need different magnitudes of stimulation for optimal performance. For instance, extroverts have a low natural level of arousal and seek stimulation, whereas introverts have a higher level of arousal and prefer low levels of stimulation.

Many years ago, I suspected that architects and interior designers are more extroverted than those in other job functions, and as such, they design workplaces based on their personal perception of what is required. However, my research on personality and collaboration (Oseland, 13b) didn't validate my hypothesis, but it did however show that architects and designers score higher on the openness scale on *The Big Five Personality Inventory*. This means that architects and designers are more open to new experiences, have a wider range of interests and fascination with novelty, plus they tend to be more creative and artistically sensitive than other disciplines. I also found that more open people value daylight and views out plus they prefer the bar/hotel, huddle room, brainstorm rooms and cafés for generating new ideas but prefer not to use meeting rooms. So, it does seem likely that the personality of architects and interior designers does affect their perception of what is required in the workplace for everyone. Therefore, it is important to provide a comprehensive design brief to the architects which explains the organisational needs and occupant requirements.

There has been much research highlighting the impact of noise on introverts, extroverts and other personality types. Much of the earlier research on the impact of noise on task performance, notably by Donald Broadbent and colleagues at Cambridge University in the 1950s, generated results in support of arousal theory. In summary, extroverts generally cope better with noisy environments, whereas introverts find noise more distracting which, in turn, hinders their performance. However, other factors also affect the response to noise such as the task being performed, attitudes, context and the source of the noise.

I have spent many years researching how to reduce noise distraction in offices; practical solutions will be presented in Chapter 4. In a survey of 2,145 respondents, across a range of organisations and countries, conducted with Paige Hodsman, we found that approximately 50% of workers consider workplace noise to adversely affect their wellbeing and increase stress (Oseland and Hodsman, 2020). Furthermore, 67% of our respondents rated the effect of noise on performance as negative, and the mean estimated impact on work performance was −6%. So, clearly, noise in offices is an issue. As reported in other studies, our survey found that introverts were more negatively affected by noise, especially if located in fully open plan environments such as a *Workplace Zoo*. We also found that the more emotionally stable and agreeable coped with noise better than their counterparts.

Furthermore, noise ratings differed significantly depending on job role. In particular, we found that the performance of data processors, analysts and researchers was more affected by noise than the other job roles combined. Data processors and analysts were found to be significantly more introverted than those in other job roles. Counter to our expectations, we found that our older respondents were less affected by noise than the younger ones. However, a higher proportion of the older respondents worked in private offices, and they may have more senior positions and autonomy. So, the age effect may be confounded by the type of workspace and role. We found overall that a higher proportion of those in private offices coped with noise better than those with allocated open plan desks. However, remote workers are less affected by noise than those in private offices and those who hot-desk in the open plan offices fare better than those at allocated desks. So, private offices are not the only solution to noise and can lead to other issues.

In my study of workplace preferences (Oseland and Catchlove, 2020), I found that extroverts had a higher preference for agile working and hot-desking, whereas introverts had a higher preference for private offices. Like introverts, those higher on the neuroticism scale also had a lower preference for hot-desking.

## Sensory processing and multisensory design

Most of us are taught the five core senses of sight (vision), hearing (auditory), smell (olfactory), taste (gustatory) and touch (tactile), whereas an article in *New Scientist* reported that we have at least 21 senses (Durie, 2005). However, seven senses seem to be popular in neuroscience, adding vestibular (movement) and proprioception (body position) to the original five. The vestibular sense is the movement and balance sense, providing information on where our head and body are in space. Proprioception involves our muscles, tendons and joints and relates to our movement and body awareness in space.

In their seminal book *Sensory Design*, Joy Monice Malnar and Frank Vodvarka (2004) explore the range of our responses to spatial constructs, highlighting how our experience and connection to a space is based on more than just vision. In particular, the sense of smell has a more direct link to the brain than other senses and can produce an emotional response based on a forgotten memory associated with a particular odour. Designing for all the senses is referred to as sensory or multisensory design. Melissa Marsh and Kristin Mueller explain that "Multisensory design stems from the idea that humans experience space in many ways – more than just through what we see" and "acknowledges that people experience and react to space in many ways, subtle and obvious, consciously and unconsciously" (Marsh and Mueller, 2017). They note that the education, retail and hospitality sectors have all begun to incorporate multisensory research into their designs, and the workplace is now starting to draw on these precedents.

Marsh and Mueller point out that a multisensory workplace design is most successful when it pays attention to and integrates all senses to give the occupants a well-rounded experience. A designer colleague once told me how the weight of an office door and the feel of the handle translated subconsciously as quality. Furthermore, focussing on just one sense can leave "blind spots" that then degrade the whole experience. For example, there is little point having an attractive looking well-planned breakout space if the waste bins are not emptied leaving odours or if the lighting is harsh or the environment is perceived as too noisy or the chairs are uncomfortable, all elements of a sociofugal space.

There is a growing field of research, relevant to the workplace, referred to broadly as sensory processing, incorporating sensory integration, sensory modulation and sensory intelligence. Tania Barney and Paige Hodsman reviewed the field and explained that "sensory processing has been described as the way in which the nervous system receives, organises, and understands, sensory stimuli from both within and outside the body, to enable a person to determine how to react to environmental demands" and "considers the neurophysiological and behavioural components of individuals in the interactions with their daily work and life environments" (Barney and Hodsman, 2020).

We all differ in sensitivity and our responsiveness to the seven (or more) senses, with our response lying along a continuum of under- to oversensitivity. Though originally developed for children, Dunn's (1997) model of sensory processing is widely used to explain sensitivity; it combines neurological thresholds with behaviour. People with a low neurological threshold are hypersensitive to stimuli and act to counteract their experience through "sensation avoiding". Conversely, those with a high threshold and corresponding hyposensitivity attempt to counteract their reduced experience through "sensation seeking". As mentioned, the level of sensitivity and, therefore, level of stimulation preferred, may vary for light, sound, odours, etc. Not being exposed to the preferred sensory level is likely to impact on wellbeing and performance. Organisations, such as Sensory Intelligence (Lombard, 2007), profile an individual's threshold for each of the senses so that an appropriate working environment can be created.

Learning, processing or communicating differences are collectively termed neurodiversity by the British Psychological Society (Weinberg and Doyle, 2017). This includes people with congenital conditions such as autism spectrum disorder, Asperger's syndrome, ADHD, dyslexia, dyspraxia and cerebral palsy. It also includes those with acquired conditions such a mental health issues, dementia, Parkinson's, multiple sclerosis and the after-effects of stroke. Kay Sargent and colleagues at HOK found that extreme hyposensitivity and hypersensitivity are prevalent in people who are neurodiverse and must be considered when designing for them (Sargent et al., 2019). Some organisations are employing neurodiverse people because they have high cognitive functioning, for example, in fields such as coding. Such organisations should take additional care in the design of their workplaces and how it affects all the senses.

The main point is to design for all the senses to enhance the experience of office workers. However, also take into account that their preferences for each of the senses will vary. At minimum, consider the impact of the workplace design on the sensory seekers and sensory avoiders.

## Inclusivity and diversity

Whilst inclusivity is not a psychological theory *per se*, it is nevertheless an important factor in workplace design that ties in with the psychology described in this chapter and the concept of designing for individual differences.

An environment specifically tailored to each individual is impractical (and probably unsuitable for new or replacement staff), but we can design for a number of groups or typologies of people. We also need to consider the full range of typologies and check to see whether the population of occupants is skewed towards one end of the range. This skewness is often considered in specification standards for specialist environments, such that the design is based on the relevant anthropometric data for the target occupants, for example, children in schools, but this is not the case in standard offices. Some individual factors to consider are noted below.

- *Personality* – Historical research, along with my own, shows differences in preferred meeting space and acceptable sound levels for different personality profiles. The modern workplace should cater for all personality types rather than focus on catering for extroverts alone. Whilst it may be impractical to design the workplace to meet each specific individual requirement, the start to the solution is to profile the team, or department, and design the main workspace to best suit the dominant personality type and activities. This is not so arduous as it first seems because certain personality types are attracted to particular jobs. Furthermore, most organisations assess personality at the recruitment stage, but then locate everyone in the same space. Accordingly, provide a calming base area for say introverted analysts, researchers or developers, but provide a stimulating base for more extroverted sales or marketing staff. However, remember that introverts will still need spaces to meet and interact, etc., whereas extroverts will occasionally need spaces for privacy or chilling, etc. Offer staff a wide choice of work-settings, away from their desk and empower them to use the spaces as and when required.
- *Neurodiversity* – Steve Maslin (2009) is a long-term evangelist for inclusive design and recognises that neurodiverse people find certain designs over-stimulating, confusing and stressful. Lighting, signage, glass, colour and art, etc. require additional consideration for this group of people. Companies employing neurodiverse people usually provide buddies and safe (quiet) areas for such staff when they feel overwhelmed and suffer anxiety. The team at HOK (Sargent et al., 2019) has created guidance on designing a neurodiverse friendly workplace. They point out that:

> The most common workplace challenges centre on the issue of sensitivity. Neurodiverse thinkers often can be over- or under-stimulated by factors in their environment such as lighting, sound, texture, smells, temperature, air quality or overall sense of security. Any comprehensive approach to designing for neurodiversity should carefully consider these experiential aspects of the work environment.

Kay Sargent and her HOK colleagues emphasise the importance of intuitive spatial organisation (layout) including clear wayfinding and spatial characteristics such as shared open spaces for socialising and semi-enclosed spaces for refuge, that is, feeling safe and secure. They also recommend providing a choice of calming environments including technology-free areas, tranquillity/focus rooms and quiet areas. Blue/green colours, natural materials and nature patterns are also recommended, "in fact many of the principles of biophilic design that have been shown to benefit neurotypical people in the workplace can also contribute to an environment that's more inclusive of neurodiversity". Some neurodiverse people may feel under-stimulated so allow for social activity and movement using connecting stairs and "chairs for bouncing, wobbling and balancing". HOK also suggest facilitating doodling/drawing in collaboration areas.

- *Age* – In my survey of noise in offices (Oseland and Hodsman, 2020), I unexpectedly found that the older occupants had less of a problem with noise than their younger colleagues. I expected younger people to be more able to adapt to noise, but that did not appear to be the case. The cynics might say that noise is not a problem for more mature occupants as they can't hear it so well. But that is not quite how the ear works and the older we are the more difficult it is to hear the speech frequencies above other sounds. It is more likely that the older respondents have more control over when and where they worked, so they are more able to choose the environment that best suits the task in hand.

  Jeremy Myerson and colleagues (Myerson, Bichard and Erlich, 2010) pointed out a decade ago that we need to design for all three generations in the workplace and not just focus on millennials.[6] The range of age groups may have different preferences for sound, lighting, layout, furniture, technology and communication media. Myerson et al. pointed out how the physiology of older workers differs from that of younger ones, and those considering designing for the generations in the workplace often assume that the elderly have more specialist requirements. However, there is growing evidence that younger people are also suffering from eyesight, hearing and posture problems (Milind et al., 2013).

  It is not just the workplace where the needs of all generations should be considered. Excessive background noise in restaurants is increasingly being recognised an unwanted noise rather than a welcomed buzz, particularly for the more mature clientele. So, also consider the design of staff restaurants and cafés in the workplace.

- *Life stage* – Probably more important than designing for age is designing for life stage. This refers to the stage in our homelife situation starting with young singles to new couples, married couples, couples with dependents, couples without dependents and older singles. AECOM (2014) suggested that Generation Z have similarities with Baby Boomers and refer to them as the "ages of freedom", including discovery, education, entertainment and exploration. In contrast, they suggest that Generation Y and Generation X are the "ages of responsibility", including children, mortgages and a fixed base. These two "ages" appear more relevant than age *per se*.

Creating working environments for those with dependents will help retain experienced, trained and talented employees. Working parents may prefer workplaces near crèche facilities, with easy access by public transport, or near parks where they can spend time with their offspring at lunchtime. Introducing flexible working hours and remote working (including working from home) will, among other benefits, support parents. Recruitment agencies, such as Ten2Two, specialise in finding part-time work for parents, in particular, mothers, who previously held senior positions but after having children can't commit to the 9–5 working hours. I once worked with a business that offered very flexible hours to their employees who were past retirement age as they wanted to retain them as mentors for their junior staff. In contrast, singles and couples without dependents are more likely to want a more social workplace than those with dependents. As such, places for interaction, café, bar, gyms and recreational facilities appeal to this group.

- *Gender* – Women returning to the office after childbirth may require discreet welfare facilities for expressing milk, etc., and those preparing for childbirth will also appreciate such areas for rest and solitude. Those undergoing the menopause, the last "workplace taboo", according to Kate Usher (2020), will require improved toilet facilities, additional locker space, quiet space for calming nerves and better temperature control.

- *Disabilities* – In the UK, we are all familiar with the term "DDA compliance", but this is a hangover from the *Disability Discrimination Act 1995*, which was superseded by the broader *Equality Act 2010*. Design requirements, such as access to buildings, for the mobile, visually and hearing impaired, are covered in Part M of the *UK Building Regulations 2010*, and similar regulations have been adopted in other countries. As the adage goes "not all disability is visible", so workplace design needs to think beyond wheelchair users. For example, provide levers instead of knobs for door handles because they are easier to use by those with limited manual dexterity, also consider different coloured and textured walls for the visually impaired to indicate changes in elevations along corridors.

There has been a very gradual move towards universal design in architecture, introduced in the 1960s, now sometimes referred to as inclusive design. Selwyn Goldsmith (1963) is associated with promoting easier access for people with disabilities, but the term universal design was coined first by architect Ronald Mace (1985). In its broadest sense, it is "the design of mainstream products and/or services that are accessible to, and usable by, as many people as reasonably possible" regardless of their age, ability or status in life (BSI, 2005). It is intended as the default position, integral and fundamental to building design rather than viewed as bolt-on solutions required to convert designs to ones suitable for minority groups of people. The seven key principles of universal design, developed by Ronald Mace and colleagues (1997), are impressive aspirational and laudable design guidelines. So, provide the same means of use for all users, avoid segregating or stigmatising any users and make the design appealing to all.

- *Culture* – The office should be welcoming to and respectful of those from all ethnic groups and cultures. Some people will require specialist facilities such as a prayer room. Too often, these are pokey afterthought spaces in the basement or part of a shared facility such as the welfare/multi-faith/first-aid room. Also consider the food offering for different cultures and provide specialist food preparation areas. Provide space to celebrate cultural holidays through food and displays. There may be other customs and traditions to consider in the workplace, determined through consultation with all cultural groups represented in the workforce.

As Neil Usher (2018a) phrased it in his pivotal book *The Elemental Workplace*: "a fantastic workplace should be fantastic for everyone" and his tenth element is "inclusion". Usher also points out that the occupants should not have to declare their personal circumstances in order for the workplace to fully accommodate them, and "the space should simply allow everyone to access, use and depart without cause for drawing undue attention, or needing to make a specific request".

Nevertheless, we are a long way off a truly inclusive society with universal design, but we must consider the design of offices if we wish to attract the full spectrum, and corresponding wider expertise and skill set, of all types of people. Putting it in statistical terms, we need to design for the range not just the average. We often see corporate businesses criticised for having an executive board of middle-aged white men, but there is a growing awareness that the workplace industry is unconsciously only designing for this group of end-users.

Let us embrace our differences, starting by recognising them and designing our workplaces to accommodate them. To achieve this, we must offer choice of a range of spatial settings and environmental conditions rather than a one-size-fits-all solution based on the assumed (or sometimes dictated) average.

Psychology demonstrates that humans are complex, and we share some common needs, but we also have varying requirements. As expected, job role and work activity affects the occupants' workplace requirements, but the inhabitants of any office vary in personality and other personal factors. Assuming a homogenous design will suit all will simply lead to dissatisfaction, discomfort, loss of performance and poorer wellbeing. Humans may be the same species, but we are all different "animals" with varying requirements. The challenge to resolving the issues with the *Workplace Zoo* is to respect individual differences and design for the wide range of requirements enabling the occupants to perform to their maximum potential. One of the biggest challenges for architects and interior designers is accommodating individual differences in dense and overcrowded open plan environments.

## Notes

1 Broady coined the phrase "architectural determinism", but he introduced it as a slight to those architects who believe in such simple deterministic principles.
2 300,000 years is an estimate, and the timeframe partly depends on how Homo sapiens are defined; it could be more like 500,000 years.
3 The Savannah Hypothesis, Theory or Principle referred to in this book is not to be confused with the related earlier, but more controversial, use of the phrase. Raymond Dart introduced the term in 1925 to refer to humans walking upright due to adapting to life on the Savannah after leaving their forest habitats.
4 Gore state that:

> Our beliefs are the basis for our strong culture, which connects Gore Associates worldwide in a common bond. We believe in the individual and each Associate's potential to help Gore grow and succeed. We also believe in the power of small teams, and through Gore's lattice structure, Associates can communicate freely ...

5 Maslow never presented his hierarchy needs as a pyramid and I prefer a ladder as each rung must be met before stepping up onto the next.
6 Arguably, we now have five generations in the workplace: Veterans/Traditionalists, Baby Boomers, Generation X, Millennials/Gen Y and Gen Z (with Generation Alpha up next).

# 4 The rise and failure of open plan

Our human needs inform how we should design and plan office spaces. The prevailing modern office solution is open plan, primarily due to the CRE industry, and occupiers', obsession with reducing space to save on property costs. There are advantages and disadvantages to modern open plan office design, but the intent of the original laudable concept became diluted and replaced by a cheap copy. A drive for efficiency rather than human needs has resulted in the *Workplace Zoo*, a high density overcrowded workspace with fewer partitions and facilities. As such, the term "open plan" has become a "dirty word", despised and dreaded by many workplace occupants and fuelled by the trade and popular press. Before discussing the pros and cons of open plan office design in detail, for context, it is worthwhile recapping on the development of the modern office and how open plan came to fruition. Juriaan van Meel (2000) and Rob Harris (2021) provide fuller accounts of the history of the office, but my pertinent highlights are presented below.

## The route to open plan

Working from home might be considered a relatively new extension of office life, but the Verulamium Museum in St Albans, UK, has a display depicting a Roman trader working from home back in AD50. The earliest examples of an actual office relate to the theologians, such as Saint Jerome (AD347–420) and Saint Augustine (AD354–430). The 15th-century paintings of 1st-century scribes typically show them working alone at a small desk surrounded by their books (Figure 4.1). Admittedly, the portrait by Antonello da Messina of Saint Jerome is a depiction, but, interestingly, it shows him sitting within a larger building at a semi-partitioned space offering some visual and acoustic privacy.

Scholars may have had a dedicated workspace, but scribes often shared a room when making copies of documents, and clerical workers across the globe often worked side by side in the same space. However, the history of the modern office with multiple staff progressed notably with the

DOI: 10.1201/9781003129974-6

*Figure 4.1* Saint Jerome and Saint Augustine (Antonello da Messina and Vittore Carpaccio, public domain, via Wikimedia Commons).

globalisation of trading. For example, early London merchants conducted their business in coffee houses or inns, but as trade became more widespread in the 16th and 17th century, then so came the introduction of merchants' houses. Initially, a place for traders to meet but gradually a place to accommodate clerks and keep records of increasing transactions – the early office. East India House, the London headquarters of the East India Company, was built in 1729 on Leadenhall Street and was possibly the first large corporate office. However, it was preceded, just, by the Old Admiralty Office, built in 1726 to handle the paperwork of navy trade routes.

Harris (2021) provides the full details of developments between the 18th and 20th centuries, particularly in London. He highlights that:

> as the office economy was growing in scale and complexity, particularly in the banking and insurance sectors, so the demands on 'paperwork' – invoices, records, lists, letters, ledgers, accounts – all grew. And alongside the paper was the proliferating office 'technology', including typewriters, telegraphs, telephones, printers and adders.

As commerce grew, so did the need for commercial offices, first with the traders, bankers and insurers and then with the clerks, counting houses, shippers, wholesalers, etc.

The introduction of the telephone and typewriter instigated the need for each employee to have their own desk. In the early 20th century, Scientific Management, known as Taylorism influenced how desks in offices were planned to maximise work efficiency and reduce costs. Elevators and revolving doors introduced at the end of the 19th century meant that multiple floors could be occupied. This all resulted in the development of space efficient high-rise open plan offices with those serried rows of desks. There were exceptions such as the Johnson Wax Headquarters in Wisconsin, designed by Frank Lloyd Wright and completed in 1939, which is a fine example of early open plan design. The building is predominantly open plan, and it features many curvilinear forms plus high ceilings with the desks laid out in small clusters, all adding to a sense of spaciousness. Nevertheless, for most European and US buildings at the time, efficiency was the key driver of office design.

In Germany, the post-war era provided opportunity for new construction and new ideas, perhaps rejecting their regimented past along with the strict order of corridors and desks. So, the Quickborner consulting group, established in 1958 by two German brothers Wolfgang and Eberhard Schnelle, introduced the concept of *Bürolandschaft* or the "landscaped office". Their principle was to create egalitarian space using irregular geometry to create a more organic, natural and humane layout, see Figure 4.2. The space was more interactive but used small clusters of desks separated with plants and curved screens to offer some privacy. An informal "break room" was also included in the space plan for leisure and coffee.

*Bürolandschaft* was adopted in principle by many organisations across Europe but gradually lost favour. This was partly due to the European economic crisis of 1970s as large well-spaced, organically planned, office areas with air-conditioning and artificial lighting were considered expensive. In addition, the large open areas resulted in some occupants complaining of a lack of visual and acoustic privacy despite the screens

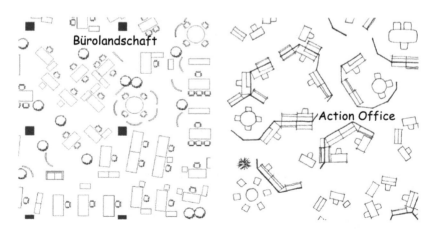

*Figure 4.2* Bürolandschaft and Action Office (Bürolandschaft by Newell Post, public domain, via Wikimedia Commons and Action Office with permission of Herman Miller Inc.).

and planting. I find this ironic as ongoing economic constraints meant that European organisations strived for more space efficient office layouts, with fewer (expensive) partitions and higher density layouts, more akin to those early Taylorist offices. Maybe the *Bürolandschaft* layout was just too natural, too disorganised, thus lacking structure as well as being space hungry or maybe managers simply worried that they had less control over their staff.

Over in the United States, Robert Propst researched office work and found that it involved much processing of information. He concluded that privacy, as well as interaction, was required for the majority of work activities. In response, Propst, with the help of George Nelson, developed *Action Office I* which was launched by Herman Miller in 1964, see Figure 4.2. The system included desks and other spaces of varying heights and screens for privacy in a fairly organic layout. Unfortunately, *Action Office I* with its tailored options was considered too expensive by the large corporates. *Action Office II* addressed this by offering a relatively easy to assemble three-sided modular panel system. Over time, the *Action Office II* was copied and degraded using cheaper materials, and the organic layout replaced with something more orthogonal, eventually resulting in the rise of the cubicle. Propst has been highly vocal and dismissive of cubicles replacing the *Action Office*: "The cubicle-izing of people in modern corporations is monolithic insanity" (Lohr, 1997), and furthermore, "The dark side of this is that not all organizations are intelligent and progressive … Lots are run by crass people who can take the same kind of equipment and create hellholes" (Abraham, 1998).

It appears that despite the compelling and admirable design concepts of Quickborner and Propst, European offices became more open plan over time, whereas the United States opted for cubicles. The conversion of *Bürolandschaft* to open plan and *Action Office* to cubicles is like a cheap pirate copy with the original replaced by an inferior product, lacking the quality and misunderstanding the intention. Offices in other countries followed the UK and Europe, depending on the sector and culture. Some Northern European countries avoided over-densification as they were protected by their Works Councils, but this often resulted in compartmentalised buildings with smaller offices accommodating several staff.

Along the way, there have been other variations in office layout, including the combi-office and the social democratic office such as Centraal Beheer's early 1970s Apeldoorn headquarters. The combi office usually consists of glass-fronted private offices arranged around a common area that includes informal meeting and breakout. Herman Hertzberger's design for Centraal Beheer is unique and comprises 56 cubic elements, each 9 × 9 m, located around a core. The elements are designed to be flexible and adaptive, and some cubes form meeting spaces and others office space, where the office layout is based on the preferences of its occupants.

Later advances in the office, such as agile and activity-based working discussed in Chapter 8, are more related to the use of the workspace rather than the design and layout. Nevertheless, as the name implies activity-based working, introduced by Veldhoen + Company in the 1990s, includes a range of work-settings that better support the range of work activities. However, apparently, it was Robert Luchetti who first introduced the concept of activity settings in the 1970s, recognising that people need multiple workplaces and different settings for different activities (Stone and Lucjetti, 1985). Office furniture manufacturers now offer furniture tailored to suit different work-settings. For example, Herman Miller's (2013) *Living Office*, including its haven, forum, cove and landing.[1]

A lesser-known space planning concept is the "free-range office". During my days working at the architectural practice Swanke Hayden Connell Architects, our Interior Design Director Nick Pell and colleagues developed what we loosely referred to as the *Free-Range Office*. Nick reminds me that "The free-range concept was primarily borne out of the need to design workplaces that are supportive and uplifting, both functionally and emotionally, within the confines of typical standard office buildings that are largely formulaic, prescriptive and constraining" (Pell, 2021). The concept, first revealed back in 2003, included a choice of work-settings (with a unique nomenclature) that could be easily reconfigured and planned off-grid to disguise the rectangular nature of buildings, see Figure 4.3.

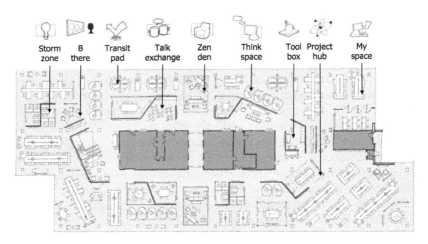

*Figure 4.3* Free-range workplace concept (courtesy of Nick Pell and Aukett Swanke Ltd.).

## Open plan versus private offices

Despite the shift to more open plan working environments, there have always been private/enclosed offices in the workplace. From the time when open plan workplaces were first introduced, private offices were provided for managers (usually overlooking the open space). Nowadays, private offices are predominantly the realm of academics, lawyers and board members, but some countries and cultures cling on to the private office more than others.

Articles warning of the perils of open plan offices are a regular occurrence in the trade and popular press. Such articles are fuelled by specific research studies which show that occupants are less satisfied after a move to open plan. The research is then extrapolated, such that the journalist claims that all open plan leads to discomfort, ill-health and degraded performance. There are five major research studies that have been favoured by the trade press for demonstrating the need for private offices.

1   *Oommen, Knowles and Zhao (2008)* – A detailed literature review that mostly focussed on psychological theories with a few field studies. In the abstract, they summarise that "employees face a multitude of problems such as the loss of privacy, loss of identity, low work productivity, various health issues, overstimulation and low job satisfaction when working in an open plan work environment", which the popular press picked up on. However, they conclude that "just because a workplace has an open plan design, it does not mean low employee productivity", and "in spite of several issues that arise with an open plan work environment, it is imperative to note that not all employees have the same problems".

Oommen, Knowles and Zhao do define open plan but, like most studies of this nature, they do not offer any detail of the specific open plan designs in their reviewed field studies. They acknowledge that those who have worked in a traditional closed office setting will have difficulty adjusting to an open plan one and that "a variety of factors have been shown to mediate between employees and open plan work environments which include job complexity and job type". The key message of the paper is for managers to better understand open plan environments and for employees to be consulted so that any potential issues are resolved.

2   *Jan Pejtersen et al. (2011)* – This Danish study found that the average reported sickness absence of 2,403 Danish workers was three days higher in open plan working environments compared with those in private offices. The Danish study is thorough, but the absenteeism rates were based on self-reporting and recall which is highly subjective. Furthermore, the authors state that "as sickness absence is a rare event, Poisson regression was used to model the number of self-reported sickness

absence days". So, it seems that absenteeism was predicted using complex log-linear (Poisson) analysis rather than it all being reported.

The Danish authors list the possible causes for increased absenteeism in open plan as: noise, viruses, ventilation, privacy and psychosocial factors. They explain that absenteeism may be related to the presence of other humans and a reduction in autonomy as "the absence of physical boundaries will increase the likelihood that co-workers and leaders will interfere with the employees' discretion and freedom to work". So, what they really found is that how the workforce is managed in open plan is more crucial than the actual design and layout.

3   *Kim and de Dear (2013)* – A re-analysis of a US survey of 42,764 respondents concluded that "our results categorically contradict the industry-accepted wisdom that open plan layout enhances communication between colleagues... This study showed that occupants' satisfaction on the interaction issue was actually higher for occupants of private offices".

The dataset is impressive, but only 6.7% of the respondents worked in "true open plan", whereas 60% worked in cubicles. The authors even report that, overall satisfaction, ease of interaction and sound privacy were better in open plan workplaces than cubicles. Kim and de Dear conclude that noise and privacy are the key variables, but their analysis clearly shows that the "amount of space" has the highest variance (explaining the difference in responses). So, density appears to be a confounding variable when exploring open plan offices. Therefore, the interpretation by the trade press is more of a commentary on cubicles rather than open plan *per se*.

4   *Gensler (2013)* – The architectural practice Gensler conducts a regular survey of office-based knowledge workers. The 2013 survey found that "only one in four U.S. workers are in optimal workplace environments. The rest are struggling to work effectively, resulting in lost productivity, innovation, and worker engagement". Despite the misinterpretation by the popular press, Gensler's report does not directly compare the responses of those in private offices with those in open plan, but there is a reference to the reduced space per person over time. Confusing issues due to increased occupational density with open plan design is a common re-occurring mistake in such research, but not all open plan environments are high density.

After Gensler's 2016 workplace survey, Janet Pogue McLaurin (2018) stated that:

> The degree of open or enclosed didn't matter in high-performing work environments. If the space was designed to function well, all

individual space types were rated as equally effective. An open plan office can be just as effective as a private one. What matters is that design aligns with employees' needs.

They found that the difference in overall effectiveness was greater between low and high functioning organisations than between private offices and open plan desking.

Following their 2019 survey, Gensler (2019) reported that their research "disproves the current narrative surrounding the open office debate" and "the research finds employees want, and expect, a great experience at work – spaces with mostly open environments combined with the right amenities and on-demand private space are the ones that deliver this best". They conclude with "one of the most important decisions companies need to make is how open they should make their office".

5  *Bernstein and Turban (2018)* – A study promoted by the *Harvard Business Review* in which the authors found that, when moving to open plan "Contrary to common belief, the volume of face-to-face interaction decreased significantly (approx. 70%) ... rather than prompting increasingly vibrant face-to-face collaboration, open architecture appeared to trigger a natural human response to socially withdraw from officemates".

To evaluate interaction, the researchers had their participants wear a sociometric badge which included a microphone, infrared sensor, accelerometer and location tracker. When conducting research, it is best practice to minimise invasiveness and avoid experimental elements that may influence the results. It is unlikely that the cumbersome sociometric badge worn by the participants did not affect the results, especially if their colleagues learned that there was a microphone attached.

The researchers reported an average of 5.8 hours of interaction per day prior to the move to open plan, which was reduced to 1.7 hours per day post-move. For most organisations, having their employees engaged in interaction for approximately 75% of their working day is not conducive to productivity; it would leave only about 2 hours per day to complete the focussed work. It could be counter-argued that the new open plan design allowed the occupants to work with less distraction and be more productive.

Furthermore, the environments studied by Bernstein and Turban were not fully described. We know that the experimental organisation decided to "completely transform the wall-bounded workspaces in its headquarters so that one entire floor was open, transparent and boundaryless". It is not clear from the paper whether the occupants were moved to a poor- or well-designed open plan workplace. However, in a later interview with *Freakonomics* (Dubner, 2018), Bernstein admitted that "Everyone was in cubicles. And then they moved to an open space that basically mimicked that, but just without the cubicle walls". In another article (Burns, 2018), Bernstein states that "there was no change in technology or overall density in the new environment, and no new collaboration or focus spaces were added". The issue may therefore be one of poorly designed space rather than open plan *per se*.

The most significant oversight, common to all studies that criticise open plan, is that the open plan working environment is not fully described. The term "open plan" is interpreted differently between studies and between countries, especially the US and UK. Furthermore, occupant feedback, collated through surveys such as the *Leesman Index* (Leesman, 2019a, 2019b), clearly demonstrates that there are poorly and well-designed open plan working environments. Furthermore, open plan environments vary by their occupational density but, due to a drive for space reduction, many have a high number of desks in the same space with few facilities. Also, some open plan spaces have no partitions or screens, whereas others may be broken up occasionally by screens, planting, storage, quiet pods and meeting spaces. On the physical design side, open plan also varies by height of desk screens, desk size, floor to ceiling heights, layout of the space, floor plate size, workstation clusters, arrangement of primary and secondary circulation routes, ratio of on-floor support spaces (breakout, meeting, refreshment, focus rooms, etc.), lighting, ventilation system, colour, artwork, branding and so on. Furthermore, and more importantly, open plan offices vary by organisational factors such as role and job function, team size, management style, sector, autonomy and responsibility, work hours, salary and reward, career path and so on. If one aspect of open plan does not work, it cannot be generalised that all open plan offices do not work.

Without fully understanding what constitutes an open plan workspace and without fully cataloguing the various elements that are included or not included in the design, any conclusion is ill-informed, and it cannot be generalised that all open plan does not work. Based on the above studies, it seems likely that density, lack of any partitions and poor design is the issue rather than the open plan concept.

Unfortunately, for many office workers, open plan design has become synonymous with poor, high-density workspace, but the overarching concept really should not be confused with such environments. Furthermore, high density is not always a bad design. Consider call centres and trading floors; these high-density working environments facilitate the core activities within them. The density facilitates a buzz and energy that is a prerequisite of such workplaces. However, these environments usually offer good facilities and support spaces, have excellent building services and, if a large floor plate, usually have higher ceilings. The mistake is made when it is assumed that such environments suit all workers and that smaller desks laid out in long and efficient serried rows are conducive to their work.

Furthermore, the research studies listed tend to ignore the transition process. Most of the studies involve a move from a private office to open plan. Moving staff to a very different new working environment will cause problems if the staff are not consulted, and the transition is not well-managed. Mixson (2019) explains that, for a successful open plan workspace, the vision needs to be fully communicated, the culture must be aligned with, acoustics should be considered and a range of spaces need to be provided. Brem (2019) points out that it is not so much the design of the space, but how open plan offices are later managed and used that causes problems.

Due to the focus of the real estate community on cost reduction, open plan is likely to continue as the primary office workspace solution – it's here to stay. In such circumstances, it is better to focus on improving the open plan environment, particularly the poorly designed ones, that *Workplace Zoo*, and better manage its use and the transition process.

## Benefits of open plan offices

The above studies focus on the negative aspects of open plan and rarely include, for balance, any positive research or published benefits. In contrast to those academic studies, case studies presented at conferences and occupant feedback surveys, including my own database and large data sets like the Leesman, often highlight the benefits of well-designed open plan offices. When referring to 1,600 workplaces with a mix of open plan and private offices, Peggie Rothe (2017a) of Leesman notes that:

> The message is clear, if rather self-evident: both open environments and more enclosed office concepts can be successful, or can fail. In the workplaces where the majority of respondents work in enclosed offices, individual employee $Lmi^2$ scores range from 46.4 to 77.2, while the range for more open concepts is 36.8 to 81.7.

Researchers at the University of Leeds (Davis, Leach and Clegg, 2011) offer a robust review of the pros and cons of open plan offices. They provide evidence that open plan can aid inter- and intra-team communication by facilitating greater communication, interaction and inter-personal relations. They also report that open plan offices "initiate and support more open and collaborative working practices, to integrate business functions, and to reflect a lack of hierarchy" and employees in a "more open office reported their organizational culture as being more innovative, less formal, providing more professional control, and fostering greater collaboration". In their discussion of the downsides of open plan, they also observe that the problems are "consistently associated with high density, open-plan offices with relatively few physical screens between staff".

Valtteri Hongisto and colleagues in Finland (2016) conducted a detailed academic study of an open plan refurbishment. They conclude "Various studies draw a negative picture of the environmental satisfaction in open plan offices. However, our quasi-field experiment provides strong evidence that such stereotypical thinking might not be constructive" and we suggest that "Environmental and job satisfaction can be improved in an open plan office if the refurbishment concerns such factors which employees have complained about, the change management is qualified and the employees are involved with the planning of the change".

Basically, Hongisto *et al.* point out that open plan involves many design and operational elements that are often overlooked in both research studies and in actual design. The complexity of open plan design, compared to a private office, may be underestimated by workplace designers, and if the drive for implementing open plan is an economic one, then it is unlikely that time and resource is allowed to tackle that complexity.

Open plan workspaces are undoubtedly borne out of a Taylorist view of work and a drive for time and space efficiency. Open plan desking offers clearer lines of sight which can aid team-working through easier interaction and aid supervision of teams. The open plan particularly supports close-knit teams, working on the same project or for the same customer, that benefit from shared (tacit) knowledge and continuous support for each other. Mixson (2019) concludes that an open plan office layout can not only improve collaboration but also spark creative thinking.

Open plan better supports teams that are involved in the same activity rather than those involved in mixed activities that may conflict with each other. In particular, a team all involved in heavy phone use will not be distracted by others on the phone and will probably benefit from the buzz of the environment. In contrast, a team whose members are all involved in complex analysis could work well together in open plan, but all will require a quiet and calming environment, reminiscent of large groups of people working independently in a library without distraction. Without good management and design, open plan will struggle to support teams where the staff are involved in conflicting activities such as some on the phone while others are involved in complex analysis. Teams involved in conflicting activities need to be separated, and spaces need to be provided to support those different activities.

In my pre-pandemic survey of office preferences (Oseland and Catchlove, 2020), I unexpectedly found that overall the respondents least preferred private offices compared to other workspaces. This surprising result was not due to a lack of managers in the sample, who seem to recognise the benefits of sitting amongst their team. However, the sample was mostly UK and Northern European-based, who favoured private offices significantly less than the Eastern European respondents. For the whole sample, there was significantly higher preference for open plan workspaces, but the highest preference overall was jointly for landscaped offices and agile working. Clearly, the sample distinguished between standard open plan environments and other versions such as a landscaped office. Furthermore, administrative staff and researchers or analysts had a higher preference for private offices and had the least preference for a landscaped office or agile working. It is worth noting that those academics, conducting research on the perils of open plan offices, may have a bias towards private offices.

Personally, I struggle with seeing the benefit of providing a building in which the workers are placed in rows of isolated boxes; they may as well work in isolation at home. In another of my surveys (Loneliness Lab, 2020), I found that those occupying private offices had higher rates of workplace loneliness than those in open plan. Loneliness is a growing issue affecting approximately 1.02 million offices workers in the UK and estimated to cost employers from £2.2 to £3.7 billion per annum (Co-op & New Economics Foundation, 2017). There are numerous ways of helping to reduce workplace loneliness through office design. For example, the space can be designed so that the occupants are more likely to mingle and "bump" into each other at building nodes, stairwells and in potential (social) interaction areas with sociopetal layouts such as café, breakout or outdoor spaces and by providing games/hobby zones. The key message is that open plan rather than private offices can help prevent workplace loneliness.

In zoos, solo animals in small cages do not fare well; this is partly due to the size of the cage but also due to their isolation. Of course, occupants of private offices can leave their door open, invite their colleagues in or choose to leave their "cage" to see others. Nevertheless, my loneliness research shows higher levels of loneliness for those in private offices, and the loneliness of senior managers, usually with their own office, is well documented in the research literature. It may be that approaching someone on an office for help is more daunting than calling out over the desk cluster, and the private office may not be adjacent to the main team desk area.

Back in the 1970s, Tom Allen found that the larger the distance between people's workspaces, the lower the frequency of their communication (Allen and Fusfeld, 1975). He found that engineers are four times as likely to communicate regularly with someone sitting 2 m apart than with those 20 m away, and he suggested that colleagues almost never communicate with those on separate floors or in separate buildings and presumably separated by walls. Allen plotted the level of communications against distance, and this became the *Allen Curve*. The *Allen Curve* still stands true despite advances in communication technology.

Placing two male animals in a single cage is also not advised as they will fight for dominance and territory. Paired offices are less common practice but can be seen in academia and law firms. However, in academia, the two who are sharing are usually not in at the same time, and in law firms, it is usually a senior partner and junior who share. Of course, it is also not advisable to place a lion and gazelle in the same cage; it is far from representative of the ratio of lions to gazelles in their natural eco-system. However, different personality types, or those involved in conflicting activities, may be forced to share a paired office or small group office, sometimes resulting in discomfort and disharmony.

I need to caveat my above view on private offices as I do not want to appear as an ambassador for open plan offices, although I am an advocate of well-designed open plan. In the UK and elsewhere around the world, open plan offices are being created that focus too much on team interaction with little regard for concentration and solitude. However, my point is that open plan should not be a "dirty word"; it is the misguided application, misuse, lack of management, poor design and low-cost execution of open plan offices that causes problems, not the original principle and intentions. In housing, open plan design infers light, airy and open well-designed and even glamorous spaces. Sadly, the term "open plan office" does not instil the same vibe, confidence and enthusiasm. So, let us not quite kill off open plan just because it has been misinterpreted and adapted to be a cheap and nasty diluted version of the original concept with all the best bits taken out.

Rather than fully dismiss private offices, I recognise that they may better suit some specialist workers. However, I am equally not convinced that private offices, at the one extreme, or misdesigned over-dense open plan workplaces, at the other extreme, suit the majority of office occupants and are another form of *Workplace Zoo*. The key is to carry out consultation, desk-top research and build the business case balancing solid evidence of the benefits with cost. Nevertheless, mostly, due to cost savings, especially in central London and other European cities where rents are ridiculously high, open plan is here to stay, and it is unlikely that we will see a shift back to corridors of private offices. The trick is to improve the open plan beyond the poorly conceived versions some occupants are now subjected to.

In my survey of office preferences (Oseland and Catchlove, 2020), I also found that, in general, the office type that people preferred was the type of office they currently occupied, in other words "we like what we know". However, it was not a clear-cut result: fewer than half (46%) of those in private offices preferred a private office, but two-thirds (66%) preferred a landscaped office and three-quarters (72%) agile working. Similar results were found for those in open plan with just 43% wanting to continue in open plan and 89% preferring a landscaped office and 80% agile working.

## Notes

1 Herman Miller describe a "landing" as an open perching spot adjacent to meeting spaces, where people can warm up before meetings and cool down after they end. My research on personality and collaboration conducted on behalf of Herman Miller (Oseland, 2012) indicated that introverts appreciate such a work-setting.
2 LMI refers to the *Leesman Index*: a 0–100 linear score of how well the workplace design and its services supports the employee in the role they are in.

# Part 2

# Solution

# 5 Introducing the *Landscaped Office*

As demonstrated in the first part of this book, many modern open plan offices fail, resulting in poor wellbeing and performance. This failure is mostly due to focussing on saving space and associated property costs rather than focussing on human needs. My proposed solution is a contemporary interpretation of the *Landscaped Office*, a revised space planning concept based on *Bürolandschaft* and other best practice. The *Landscaped Office* is better suited to office inhabitants than the *Workplace Zoo*, that unpopular, poorly designed, overcrowded and misconstrued version of open plan.

## A (re)emerging workplace solution

My workplace consulting experience and research revealed that modern open plan offices are predominantly associated with poorly designed high-density environments with few partitions. Densely occupied offices crammed with rows of desks, little support space, over-loaded building services, poor noise control, etc., will undoubtedly lead to poor satisfaction and possibly a drop in business performance. I have even overheard workplace consultants proclaiming that good open plan working environments take up more space than private offices. High density and space efficiency were not the primary focus or the intention of the early open plan workspaces. Both *Bürolandschaft* and *Action Office I*, more prevalent in the 1950s–1960s, included an array of work-settings plus interesting and somewhat spacious layouts. Workplace design is not a simple dichotomy of private offices versus open plan, there is a whole array of different types of office space. To illustrate the point, Neil Usher a.k.a. blogger Workessence created a taxonomy of 14 office typologies (2018b).

Poor open plan may be a result of high density and insufficient partitions/screens, but Figure 5.1 illustrates how the variation in density and

DOI: 10.1201/9781003129974-8

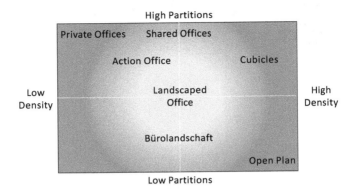

*Figure 5.1* Poor open plan is mostly due to high density and low partitions.

partitions can be applied to all office spaces. For example, open plan, with high density layouts and low screening (in both numbers and height), is diametrically opposite private offices, which are low density with high partitions. *Bürolandschaft* was quite open with few partitions, other than plants and the occasional screen, whereas *Action Office* introduced high screens between clusters of one or two desks.

Like most things, the solutions and scenarios at extreme ends of a range tend not to be the best and the optimal solution lies in the middle. Whether a fully open plan environment with no private spaces or a workspace of just private offices, neither is likely to accommodate a range of individual human needs or suit the needs of the dynamic modern organisation. While the *Workplace Zoo* predominantly relates to high density workplaces, it could easily be created through dark corridors of private offices where the inhabitants are isolated for the majority of their working day.

My proposed optimal workspace solution is a return to the *Landscaped Office*. This is by no means a new concept, but it is one in grave need of reviving and revising. *Bürolandschaft* literally translates to "office landscape", but the phrase *Landscaped Office* is used interchangeably and has been for some time. When describing the *Landscaped Office*, Ferreira, Erasmus and Groenewald (2003) explain that "The biggest difference between this approach and the open plan office design is that the workstations are not all positioned facing the same direction". Similarly, the *Dictionary of Construction, Surveying, and Civil Engineering* (Gorse, Johnston and Pritchard, 2012) defines the *Landscaped Office* simply as "An open office where the furniture, screens, equipment, plants, and decorations can be moved around to the desired layout". My version of the *Landscaped Office* is primarily a revival and contemporary interpretation of *Bürolandschaft* with elements of *Action Office* and inspiration from the *Free-Range Office* and other workplace concepts.[1] My understanding of the primary differences between *Bürolandschaft*, the typical modern version of open plan and the new *Landscaped Office* are clarified in Table 5.1.

The *Landscaped Office* is technically a type of open plan office, but it is so much more than the modern-day open plan workplace, it is "open plan plus". As with *Bürolandschaft*, there is much more emphasis on the spacing of desks, applying the optimal occupational density, in order to meet human psychological needs and help reduce noise. The layout of desks with distinct zones, activity areas and clear circulation routes is likely to be more organised than the original *Bürolandschaft*,[2] yet more organic than modern open plan offices. Well-planned circulation helps reduce distraction in the

*Table 5.1* Primary differences between different workplace strategies

| Design and planning features | Bürolandschaft | Modern open plan | Landscaped office |
|---|---|---|---|
| Occupational density | Low | High | Low |
| Partitions/screens | Some | Minimal | Numerous |
| Openness | Medium/high | High | Low/medium |
| Space plan and layout | Haphazard* | Orthogonal | Organic |
| Desk clusters | Low (1–2 desks) | High (6–12 desks) | Medium (2–6 desks) |
| Activity zones | Some | Minimal | Multiple |
| Team zones | Mostly | Mostly | Varies |
| Choice of work-settings | Some | Minimal | High |
| Flexible furniture | Some | Minimal | Numerous |
| Private offices | None | Some | None |
| Desk-sharing/hot-desking | None | None | Optional |
| Standard desk screen height | None (0 mm) | None/low (0/1,100 mm) | Medium (1,300–1,500 mm) |
| Standard desk size | 1,600–2,000 mm | 1,200–1,400 mm | 1,600–1,800 mm |
| Personal environmental control | Not usually | Not usually | Preferred |
| Biophilic design principles | Some | Not usually | Definitely |

* Appears unstructured but based on team interaction.

main desk areas but also enables communication and interaction away from the desk.

In addition to consideration of the two-dimensional horizontal space, the vertical plane, third dimension, is also key to the *Landscaped Office*. The space is better broken up with different heights of semi-partitions using planting, hanging screens, whiteboards, interactive display screens, open bookshelves, etc. Pub designers are particularly good at using screens to create semi-private nooks and crannies in an otherwise open space. Such spaces also have a different look and feel, a different ambience and aesthetic using a range of coloured lighting, wall colours and materials. The *Landscaped Office* has a fuller array of work-settings than the desks and meeting rooms of standard open plan environments, such as focus pods, 1:1 rooms, collaboration booths, informal meeting spaces and breakout areas, detailed in the next section. The work-settings can also be used to break up the clusters of desks creating smaller semi-open more private (team) zones.

## Humanising the office

The modern office should be built on the psychological, physiological and personal needs of the workforce rather than be based purely on organisational needs or, even worse, driven by a focus on reduced costs. In previous chapters, I introduced some psychology theories and associated research relevant to office layout and design. I have incorporated the psychology, along with my personal practical experience gained through projects, into a set of 12 broad workplace design principles that do not just apply to the *Landscaped Office* but are relevant to any human-centred office design; I call these my "c-words for the C-suite".[3]

1  *Choice* – Offering people choice of where, when and how they work is fundamental to performance and wellbeing. The *Landscaped Office*, with a rich menu of work-settings on offer as and when required, is more likely to meet our individual needs and inquisitive nature. It is important to ensure that the focus is not just on space reduction and provide a choice of work-settings. Nevertheless, the primary choice of a small percentage of some workers may be to occupy the same desk from 9 am to 5 pm for five days per week rather than preferring a more agile option. As Neil Usher (2018) points out, "a choice of well-specified, well-designed, and well-arranged work-settings is often the answer to so many workplace issues".

2  *Collaboration* – True collaboration is when two or more people come together and create something that they could not produce alone.

Designing spaces to bring people together and enhance collaboration has been a firm favourite with interior designers for some time. Clearly, teamwork and collaboration are important in many industries and can lead to new innovative products, cross-selling and faster time to market. Design can indeed facilitate collaboration, but it first requires the organisational factors to be in place. Such factors include the right (knowledge sharing) culture, appropriate management, rewards for collaboration and, most importantly, mutual trust and respect. Trust is a prerequisite of collaboration, and trust can be built through social interaction. Furthermore, we are social animals that desire interaction, telling stories and sharing food (yearned for more by extroverts than introverts). Mingling spaces, interaction nodes, breakout spaces, games areas, etc., all contribute to building a collaborative working environment. However, we also need to acknowledge that some roles and work tasks do not actually benefit from continuous collaboration.

3  *Concentration* – Working environments need to provide spaces for concentration as well as for collaboration and creativity. High proportions of work time continue to be spent carrying out detailed or complex tasks, so we will need areas for focus and concentration. Such areas may be favoured by the more introverted, who may also be attracted to tasks requiring extended periods of concentration. In open plan environments, we may need to offer alternative work-settings such as focus pods or quiet zones (free from phones and chatting) or allow working outside the office, such as the home or third spaces like a library. Working from home supports focussed work for some, but others do not have privacy at home and can be more easily distracted.

4  *Creativity* – This is another hot topic with designers. Providing a stimulating, colourful, quirky, fun and buzzy space sometimes contributes towards creativity as the space sets the tone for less formal meetings. However, the creative process not only involves people coming together to share and test ideas, but it is followed by long bouts of time in solitude thinking through and developing those ideas. Research shows that taking time out (from the desk/office) also assists creativity. In particular, taking solitude in natural environments by going for a walk offers a setting for "non-taxing involuntary attention", which helps us solve problems and progress. The workplace should incorporate nature by bringing planting indoors or providing a semi-covered terrace or garden area.

5  *Comfort* – It is important to provide for the hygiene factors. Based on Herzberg's (1959) *Two-Factor Theory*, if the hygiene factors are not right, then performance will be degraded (as opposed the motivational

factors, such as reward, which enhance performance). The challenge is providing comfortable environments when a group of individuals share the same space. Noise and temperature are the biggest problems in office design and the greatest challenge now facing interior designers and building service engineers. Increased choice and control improve comfort.

6  *Control* – We have personal preferences for temperature, light, noise, etc., but control of environmental conditions is often ignored. This is largely due to the difficulty of offering personal control in open plan environments. However, unassigned desks and alternative work-settings allow people to move around and choose spaces that offer the conditions that best suit them. Alternatively, offer control via environ-mentally responsive workstations or other such products. Control may also refer to organisational factors such as workload and the ability to reduce stress through better planning and the choice of when to work.

7  *Connectivity* – This is crucial both within and outside the office, re-gardless of whether in a co-located or virtual team. I recall seeing an adaption of *Maslow's Hierarchy of Needs* showing Wi-Fi as the most fundamental human need, and it is now the key technology enabler rather than the hardware. Of course, connectivity also refers to human connection. Humans are social animals, and some personality types thrive on social interaction. I find that, after working from home for a few days, I get "cabin fever" and actively seek human interaction. Workplace loneliness is a growing problem, and it has higher preva-lence in those who work from home continuously.

8  *Confidentiality* – This may be related to personal private matters or sensitive information regarding our colleagues. This does not trans-late to private offices or assigned segregated areas set aside for super-confidential teams, but spaces offering good visual and acoustic privacy are required occasionally. In future, spaces may be required for creating and recording on-line/virtual material.

9  *Contemplation* – As well as being social animals, we sometimes just need a place to chill, relax, reflect and contemplate in solitude. Areas are required for gathering our thoughts and reenergising. Such a place is calming with greenery and views out or with an outside terrace or garden; this is also good for creativity.

10  *Care* – Care as in wellbeing, health, active design, mindfulness, etc., are all very on-trend. It is great to see the recent focus on employee wel-fare, even if (cynically) it might extend their time at work or result in cheaper healthcare cover. Whilst the CRE industry, particularly, facili-ties managers and workplace consultants acknowledge the importance

of wellbeing, the ongoing fixation with cost reduction means that de-sign features that may enhance wellbeing are not always implemented.

11   *Co-location* – The co-location, or co-presence, of teams is still impor-tant to most organisations but more of a challenge in the post-pandemic world of remote working. Alternative workspaces like meeting spaces and project rooms are required; nevertheless, teams need to occasion-ally congregate to work in parallel at desk clusters. This also improves connectivity and helps build trust and loyalty.

12   *Cost* – Unfortunately, cost will remain a key metric in the workplace industry. As discussed previously, we can measure productivity (it's just difficult), and using performance metrics will lead to better, more in-formed, more human-centred, workplace solutions. At the moment, Finance Directors are blinkered by their faith in cost data (or rather lack of faith in other metrics), in other words "cost is king". The good news is that high potential cost savings from office moves at least allow investment in new and better workspaces.

There is a 13th principle, but it is more related to implementation than design.

13   *Co-creation* – Co-creation refers to allowing employees to have some input to the design from a range of feasible options. It will help the transition to a new space as well as create some alternative, interesting and eclectic interior designs. Change management is critical in the success of implementing a successful workplace, especially new strat-egies such as agile working environments. This involves communica-tion, consultation, cooperation and co-creation with those migrating to the new workspace.

## Accommodating human needs

The *Landscaped Office* accommodates the psychological and physiological needs that are often overlooked in other design strategies, particularly the modern interpretation of open plan.

Straightish lines exist in nature, but right angles do not except at the micro level such as in crystals. In the Müller-Lyer illusion, shown earlier in Figure 3.2, the line with arrow heads is usually perceived shorter than the line with arrow tails. However, the perception can vary, and back in 1901, it was found that the indigenous tribe's folk on Australia's Murray Island were not fooled by the illusion. This is possibly because the illusion relies on the arrow heads representing the inside corners of a room and the arrow tails looking more like the outside corners of a building (also shown in Figure 3.2), but the islanders at the turn of the 20th century did not inhabit such rectilinear built environments. The finding indicates that orthogonal layouts are probably not our natural preference – that came later when focusing more on work and space efficiency.

Consequently, the *Landscaped Office* is less orthogonal than typical open plan but mimics the complexity and order encountered in nature. This will appeal to our innate affinity to nature, that is, biophilic needs, rather than adhere to the Taylorist (1911) style layouts and seas of desks in rows. The *Landscaped Office* may also include alternative desk shapes and configurations to bench desking to add to the more natural feel. The 120° desk is quite popular with occupants, but less so with those focussed on efficiency and, incidentally, can be planned to look like the biomorphic form of honeycombs.

Back in Chapter 3, I mentioned Osmond's (1957) sociofugal and sociopetal spaces – the *Landscaped Office* may have both, depending on the purpose of the space. For example, a simple application of Osmond's theory is how a room is laid out for different types of meetings. Seating arranged in circles and with people facing each other is better for interaction than long boardroom tables – King Arthur's round table is an historic example. In contrast, theatre style seating with rows of seats in lines, or preferably semicircles, is better for didactic presentations. Breakout spaces that do not offer some level of privacy, refreshments, comfortable seating or a pleasant design may discourage interaction and be considered sociofugal. In the main desk area, depending on the work to be carried out by a team, the desks may be arranged so that people face each other, with lower barriers for interaction or placed so people face away from each other with higher partitions to minimise interaction, see Figure 5.2. The latter may be incorporated into a quiet working zone. People with a lower sensory threshold may prefer sociofugal spaces that minimise interaction and will prefer a corner location to the middle of the space.

A practical outcome of Altman's (1975) *Privacy Regulation Theory* is that building occupants require a means of controlling their required level of privacy depending on factors such as their personality and the task in hand. This is particularly difficult in open plan environments but can be overcome by a choice of work-settings offered in the *Landscaped Office*. As already noted, a sociofugal desk arrangement will better suit those who spend all day concentrating or do not wish to be distracted by colleagues, but the downside is that it will restrict interaction. An alternative solution is to offer other sociofugal spaces for staff to go to that restrict interaction, for example, quiet pods, focus rooms and quiet zones. For teams who mostly

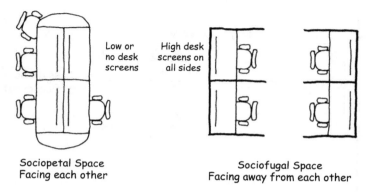

Low or
no desk
screens

High desk
screens on
all sides

Sociopetal Space
Facing each other

Sociofugal Space
Facing away from each other

*Figure 5.2* Sociopetal and sociofugal desk clusters.

conduct focussed work, provide nearby sociopetal spaces that encourage in-teraction, for example, collaboration, social and breakout spaces.

Privacy is not determined by physical factors alone but can be achieved through behaviour and agreed social etiquette. Counter-intuitively, a busy café or breakout space away from the team area can provide privacy due to the lower likelihood of interaction. In contrast, prolonged bouts of work away from colleagues, for example, at home or travelling on business, can result in isolation and loneliness for some people, so spaces in the office for social interaction are also key.

In modern open plan offices, in order to increase space efficiency, it is quite common to see desks grouped in clusters of 8, 10 or even 12, that is, four, five or six desks in each row. Desk clusters should marry up with the ideal team size so be around 8–12 desks. However, the *Landscaped Office* embraces clusters of two and four desks. This is because a cluster of more than four desks is less practical as it (i) impacts on personal space as someone will be sat between two others, (ii) increases noise as more people are sitting nearby, (iii) causes discomfort from being overlooked from behind, as per evolutionary psychology, and (iv) heightens visual distraction as the person at the end of the cluster may need to pass and walk behind three or more other staff when leaving their desk; move-ment is especially distracting as our peripheral vision is more sensitive to movement (see Figure 5.3).

So, for teams of 12, the *Landscaped Office* proposes three clusters of four desks or possibly two clusters of six desks. Providing more circula-tion between desks reduces the density of the layout. Whilst this may be

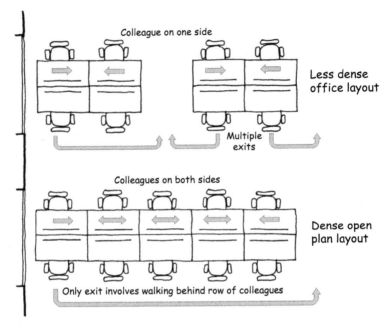

*Figure 5.3* Circulation around desk clusters.

considered space inefficient, it is worth remembering that densities have increased over time, and the laxing of it will help reduce noise distraction and cross-infection. As we like daylight and views outside, the desks should be adjacent to windows rather than in the deep plan part of the office or central core areas. Well-spaced meeting rooms with windows and daylight are also preferred to the dark and dingy ones.

For cost efficiency, office floorplates in the city are often large and deep plan with 400–600 desks. However, Dunbar's Number (1992) implies that these floorplates are beyond the human scale; therefore, it is better for the *Landscaped Office* to be located on smaller floorplates of 150 desks or for the floorplate to be broken up into smaller more comprehensible zones, that is, departmental communities or business zones.

The zones of the *Landscaped Office* may also reflect other requirements. Stimulating (buzzy) and calming (or quiet or silent) zones may be created depending on the task being carried out and the predominant personality types. whilst a base zone may be stimulating, for sales or extroverts, etc., the occupants of those zones will also occasionally require more calming environments or spaces where they can be alone. In contrast, those in a calming base zone will occasionally need nearby spaces to meet, interact, socialise and let off steam.

A calming zone can be created using subdued lighting and colours, lots of sound absorption and high desk screens. In contrast, bright, varied, coloured lighting, rich colours and furnishings, colourful art and more reflective (reverberant) finishes all help create a stimulating environment.

The design and ambience of the alternative work-settings in the *Landscaped Office* will also vary. Based on Barker's (1968) behavioural settings, using visual cues can set the scene and tone for how the space is used. A quiet space may have darker furniture and bookshelves to reinforce the behavioural setting of a library or reading room. Meeting areas with a more domestic feel and compact comfy seating will engender a different tone to the meeting compared to one in a formal meeting room with a large rectangular table and people sat opposite each other. Brightly lit and coloured spaces with few standard desks but with funky furniture, plenty of mingling and social space and writing surfaces imply a setting for interaction, innovation and creativity.

Hall's (1963) *Proxemic Framework* includes his "social distance" for interaction with business acquaintances and is estimated to be 1.2–2.1 m. However, in the typical office, it is now quite common to find desks 1,400 mm wide, and this is becoming the industry norm. Desk sizes have reduced over the years, I recall my 2 × 1 m desk at an architectural practice, but then, 1,600 mm became more common before the workplace industry settled on 1,400 mm. Recently, I have worked with clients who already use 1,200 mm desks and another proposing just 1 m desks for some of their staff; for the latter, the main activity is talking to patients on the phone and then writing up notes. These smaller desks will not only be perceived as an intrusion of personal space causing discomfort but through proximity will generate more noise, distraction and potentially cross-infection. The *Landscaped Office* calls for a return to 1,600 mm, or preferably 1,800 mm, desks.

This chapter introduced my recommended workspace solution, the revived and revised *Landscaped Office*, along with some high-level planning principles based on psychological needs. Chapter 7 provides more detail on the components of the *Landscaped Office* and how to space plan them. However, it is also necessary to ensure the environmental conditions, facilities and designs of the *Landscaped Office* support the biophilic and physiological needs of its inhabitants.

## Notes

1 Interestingly, Frank Duffy introduced the UK to *Bürolandschaft* in 1966 before going on to introduce more agile-type workplaces. I consider Frank to be the originator of the workplace consulting profession (the one I belong to), so *Landscaped Office* is also a nod to his legacy.

2 Whilst the *Bürolandschaft* space plans look somewhat haphazard, they were determined from detailed surveys of the level of interaction between workers and departments, along with the desk space required based on role and technology.

3 C-suite refers to high-level company executives who usually have a C on their title: Chief Executive Officer (CEO), Chief Financial Officer (CFO), Chief Operating Officer (COO) and Chief Information Officer (CIO). Of course, the principles are also relevant to key stakeholders not in the C-Suite, but I like the alliteration.

# 6   It's a jungle in there

Biophilia is "our innate affinity to nature", and as such, we yearn for nature, and it unconsciously affects our wellbeing and performance. Authors such as Stephen Kellert and Bill Browning have documented evidence on the benefits of introducing nature into the workplace, referred to as "biophilic design". The term biophilia has been around for at least 40 years, introduced by Erich Fromm and popularised by Edward O. Wilson (1984). In contrast, biophilic design has only been adopted in mainstream design for the past 10 years or so.

Duncan Young, who introduced biophilic design into Lendlease's own offices and their developments, told me, "Many of our everyday actions are shaped by our physical, virtual and social environments. We spend a significant amount of time at work so workspaces can help our health and wellbeing so we can thrive not just survive". However, our innate human requirements are amiss in most modern offices despite research showing that they can increase job satisfaction, improve creativity and performance, alleviate stress and reduce absenteeism.

One simplistic interpretation of biophilic design by the CRE industry is the introduction of copious plants into the office. However, biophilic design is so much broader than indoor planting and landscaping and includes all aspects of the natural world such as natural ventilation, natural light and natural soundscapes. Indeed, Kellert (2015) explicitly says:

> Biophilic design depends on repeated and sustained engagement with nature. An occasional, transient, or isolated experience of nature exerts only superficial and fleeting effects on people, and … Exposures to nature within a disconnected space – such as an isolated plant or an out of context picture or a natural material at variance with other dominant spatial features – is *not* effective biophilic design.

DOI: 10.1201/9781003129974-9

## Evolutionary psychology

It is well known that human physiology evolved to survive and thrive on the African Savannah, but evolutionary psychologists believe that the same applies to the functioning of the brain. In the last 100 years, we have migrated to the modern office, and whilst we may have adapted to this relatively new environment, the brain's evolution lags to those earlier times. The field of evolutionary psychology has helped us better understand our preferred indoor environmental conditions, directly inferred from our past of living and surviving in natural environments. The following guidance elucidates on how to move forward from the *Workplace Zoo* to create a workspace more in-line with innate human needs, thus refining the design and layout of the *Landscaped Office*.

- *Daylight* – Daylight is important for regulating sleep patterns. Access to daylight allows us to tell the passing of time during the day and the year. Consider how discombobulating it is to go to the cinema during the day and come out into the daylight. Buildings with heavily tinted windows or deep plan buildings where desks are positioned away from the windows or offices where the window blinds are continuously closed can also be disorientating.
- *Views* – Humans favour views out, preferably of nature, which harks back to a time when we were hunters but also hunted. Primitive humans would sit with their back to a tree or rock for protection and peruse the vista. Consequently, we like views to the outside and prefer not to be overlooked from behind, and we don't like sitting with our backs to a busy walkway or corridor.
- *Temperature* – We all have personal preferences for temperature and depending on our activity, and personal factors, we may like it warmer or cooler than others. Research has also shown that we prefer slight fluctuations in temperature rather than the steady-state ones found in air-conditioned boxes. In nature, if the temperature does not suit us, then we would move to a different space, in the sunshine or in the shade, or we might light a fire or adjust our clothing – such adaptations are not always allowed in the office.
- *Air movement* – Many people prefer natural ventilation to air-conditioning. On a warm day, a gentle breeze on bare arms is quite appealing rather than a down-draught on the neck. Gentle air movement provides a direct feedback loop and informs us that a space is being cooled – part of the reason why windows, and fans, are preferred to

mechanical ventilation. If natural ventilation is not possible perhaps provide sheltered outdoor spaces for relaxation and occasional meetings. Cloisters, the covered walkways around courtyards often found in traditional British universities, allow walking in the fresh air with a view of the gardens even when it is raining.

- *Sound* – A slight breeze will cause leaves to rustle or water to lap on the shore – both familiar and natural sounds. These and other natural sounds, such as birdsong and running water, are more pleasing and relaxing than mechanical sounds. Experiments have found that such sounds can reenergise workers and improve their performance, so are particularly useful in breakout areas and chillout zones. In nature, no sound signals danger from fire, poor weather or a predator, so we are more comfortable with background sound levels of around the 35 dB found outdoors in rural settings rather than complete silence.

Evolutionary psychology provides a better understanding of our preferred environmental conditions. In addition, our preferred type of different office spaces can also be inferred from the Savannah Hypothesis.

- *Inquisitive* – Desmond Morris (1967) notes that "All animals have a strong exploratory urge, but for some it is more crucial than others", and this is partly driven by the need for food. He continues that the opportunists, like primitive humans, "are never sure where their next meal may be coming from, and they have to know every nook and cranny, test every possibility … have a constantly high level of curiosity". Morris proposes that our curiosity and exploratory urge partly resulted in language and creativity through music, dance, art and writing. As we are naturally inquisitive animals, we like to explore rather than be inactive and sitting for too long. So, provide the opportunity for moving around and build interest into the building with alternative routes, unusual shapes and layouts, nooks and crannies, plus arts and points of interest, etc. The latter "landmarks" also help with wayfinding.
- *Movement* – It is most probable that our musculoskeletal system evolved to support hunting and living on the African Savannah and therefore it could be argued that continuous sedentary activity takes its toll. Prolonged sitting at a computer results in poor posture and absenteeism due to neck muscle (particularly the *levator scapulae*) and lower back pain along with repetitive strain injury. Furthermore,

standing in the afternoon was found to reduce blood glucose levels and the risk of diabetes and cardiometabolic diseases (Buckley et al., 2014). A good ergonomic chair and desk set-up are vital in alleviating back and neck pain but also provide the opportunity for staff to stand up and move around by (i) providing sit-stand desks, (ii) including stand-up meeting tables, (iii) encouraging movement around the office to go to meeting spaces, breakout and other facilities, (iv) enticing people to move by creating intriguing spaces for exploration and (v) providing well-designed easily accessible stairwells between floors. These are all basic elements of "active design".

- *Socialising* – Humans are social animals; we love to meet and share stories. Social interaction builds trust and is a precursor for true collaboration, and it also enhances connectedness to colleagues and the organisation. So, "workspace", both indoors and outdoors, is required for socialising as well as for more traditional work activities. We particularly like to share stories with our colleagues over food and drink – harking back to the hearth mentality when storytelling was our predominant means of sharing knowledge. Recreating the hearth with a gas fireplace (inside or outside), inglenook and surrounding seating will enhance the effect. An under-utilised breakout space can be boosted by providing good coffee and biscuits (Figure 6.1)!

*Figure 6.1* Breakout space with refreshments, daylight and planting at Lendlease.

- *Contemplative* – Although we are social animals, we are also contemplative and require spaces for occasional solitude, solace and privacy. Providing more natural spaces will facilitate restoration and help reenergise the weary worker. Some organisations have included art, soundscapes and even an aquarium in their chillout zones. Subdued lighting, comfortable furniture and semi-private spaces, for example, using nooks and crannies, also helps.
- *Animal affinity* – Desmond Morris proposes that our liking of and taking comfort from stroking animals relates back to our own days of social grooming. He continues by highlighting how throughout history, we have revered animals, used them as symbols of strength in sports, copied their markings in fashion and, of course, formed a symbiotic relationship with some animals resulting in befriending and domesticating them. We feel compassion to animals, some of us are keen observers of wildlife, for example, watching and feeding birds and other indigenous animals. Others seek pleasure in hunting our wildlife for pleasure, but generally, a lack of empathy for animals is considered symptomatic of sociopathic behaviour. Some organisations have allowed their workforce to bring their pets into the office, but this usually relates only to dogs. Sedum rooftops, or a simple bird feeder, will attract birds, and an aquarium is a fascinating addition to an office.
- *Greenery* – The claim that plants not only convert $CO_2$ to oxygen but also remove pollutants is well documented, with certain plants being better at removing specific pollutants, and the seminal research in this area is that of NASA (Wolverton, Douglas and Bounds, 1989). However, one of the researchers has since claimed that it's impossible to guess how many plants might be needed; nevertheless, he recommends at least two "good sized" plants per 10 m$^2$. In her review of the NASA study, Claudio (2011) concluded that plants can indeed remove toxins from the air under laboratory conditions, but in the real world, the notion that incorporating a few plants can purify the air has little scientific evidence. My concern is that the more frugal side of the CRE industry may consider replacing good ventilation with a few potted plants.

  Having criticised using potted plants for air filtration, it is evident that they have psychological benefits (obliging our innate needs), and research shows that they can help improve creativity and problem-solving. Plants are part of our natural environment, so, where practical, introduce plants into the office (Figure 6.2 and 6.3). Preferably use ones maintained by a professional supplier but note that staff may also want to bring in their own plants. Plants also have some acoustic properties and can mitigate noise issues. For example, moss

*Figure 6.2* Creating a chill-out zone and outdoor landscape indoors (image courtesy of Saint-Gobain Ecophon).

*Figure 6.3* An office floor flooded with plants at Lendlease.

walls provide some sound absorption and most larger plants, including broad leaf ones, bamboo and grasses, can create sound diffusion breaking up the sound waves as well as providing some visual privacy. Allowing staff to grow their own vegetables in outdoor containers or small hydroponic systems appeals to some. As well as indoor plants consider the outside landscaping and provide outdoor terrace seating. In contrast, it may be better to avoid overly fragrant flowering plants, as they would not be welcomed by those with olfactory hypersensitivity or allergies.

## Biophilic design

In the introduction to *Biophilic Design: The Theory, Science and Practice of Bringing Buildings to Life*, Stephen Kellert identified two dimensions of biophilic design with six biophilic design elements and 70 corresponding biophilic design attributes. The list was later reduced to a more practical three dimensions and 24 attributes, see Figure 6.4 (Kellert and Calabrese, 2015). At a similar time, Bill Browning and colleagues proposed 14 biophilic design principles categorised under three broad patterns of nature, see Figure 6.5 (Browning, Ryan and Clancy, 2014).

There is some overlap between the two sets of design principles, and some have already been discussed in this chapter. However, there are many

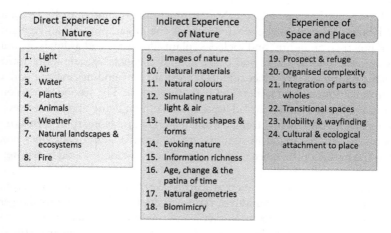

| Direct Experience of Nature | Indirect Experience of Nature | Experience of Space and Place |
|---|---|---|
| 1. Light | 9. Images of nature | 19. Prospect & refuge |
| 2. Air | 10. Natural materials | 20. Organised complexity |
| 3. Water | 11. Natural colours | 21. Integration of parts to wholes |
| 4. Plants | 12. Simulating natural light & air | 22. Transitional spaces |
| 5. Animals | 13. Naturalistic shapes & forms | 23. Mobility & wayfinding |
| 6. Weather | 14. Evoking nature | 24. Cultural & ecological attachment to place |
| 7. Natural landscapes & ecosystems | 15. Information richness | |
| 8. Fire | 16. Age, change & the patina of time | |
| | 17. Natural geometries | |
| | 18. Biomimicry | |

*Figure 6.4* Biophilic design attributes proposed by Kellert and Calabrese.

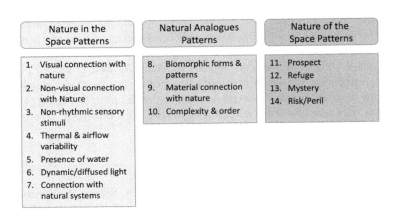

*Figure 6.5* Biophilic design principles proposed by Browning, Ryan and Clancy.

other noteworthy biophilic design principles relevant to the *Landscaped Office*. The first dimension set out by Kellert and Calabrese is the direct experience of nature, which refers to actual contact with environmental features in the built environment such as natural light, water and plants. Their second dimension, the indirect experience of nature, relates to images of nature and patterns found in nature, such as natural materials and artwork. Their third dimension refers to spatial features that are characteristic of the natural environment.

Browning, Ryan and Clancy's first dimension is similar to Kellert and Calabrese: "Nature in the space addresses the direct, physical and ephemeral presence of nature" and includes "plant life, water and animals, as well as breezes, sounds, scents and other natural elements". Examples of introducing nature into the office and its immediate surroundings include potted plants, flowerbeds, bird feeders and butterfly gardens. The second dimension, natural analogues patterns, "addresses organic, non-living and indirect evocations of nature" such as using natural materials and biomorphic forms that represent patterns found in nature. Finally, Browning and colleagues note that the nature of the space refers to spatial configurations found in nature:

> This includes our innate and learned desire to be able to see beyond our immediate surroundings, our fascination with the slightly dangerous or unknown; obscured views and revelatory moments; and sometimes even phobia inducing properties when they include a trusted element of safety.

Below are some further specific biophilic design principles, introduced by Browning, Ryan and Clancy plus Kellert and Calabrese, that contribute to the look and feel of the contemporary *Landscaped Office*.

- *Patterns* – Natural patterns and biomorphic shapes can be introduced into the workplace in various ways. The most obvious is through using natural materials such as timber and stone. Natural materials add a certain gravitas to a space, whether used on the floor or wall finishes or for furniture items. Desks with wooden surfaces have a warmth beyond the common grey or white laminated ones. It is appreciated that a wooden veneer may not be feasible for all desks, and a wood-like laminate version may be required, but real timber furniture can be used sparingly in the work-settings such as meeting and breakout spaces. There is a subtle difference between taking inspiration from nature in design and using artificial materials, such as the plastic planting often found around offices, gathering dust.
  Browning and colleagues observe that:

  Objects, materials, colours, shapes, sequences and patterns found in nature, manifest as artwork, ornamentation, furniture, décor, and textiles in the built environment. Mimicry of shells and leaves, furniture with organic shapes, and natural materials that have been processed or extensively altered (e.g. wood planks, granite tabletops), each provide an indirect connection with nature: while they are real, they are only analogous of the items in their 'natural' state.

  Fabrics can also represent nature either by mimicking animal and other natural patterns or through colour, for example, autumnal or coastal shades.
  Browning et al. also refer to complexity and order, which is the "rich sensory information that adheres to a spatial hierarchy similar to those encountered in nature". Similarly, Kellert and Calabrese refer to natural geometries or mathematical properties commonly encountered in nature such as fractals and the Fibonacci sequence seen in spirals and helixes. Such shapes can be represented in artwork and wall patterns (Figure 6.6), and most people love a spiral staircase other than workplace health and safety who consider them both a trip hazard and not inclusive. Arches, vaults and domes resemble forms found in nature such as beehives, nest-like structures, shells and cliffs. Replicating some natural complex patterns is good, but too much can also induce

*Figure 6.6* Tree patterns on meeting rooms and plants.

stress. Kellert and Calabrese note that "People covet complexity in both natural and human settings, which signify places rich in options and opportunities. Yet, excessive complexity is often confusing and chaotic".

- *Colour* – Kellert and Calabrese specifically highlight the use of natural colours and recommend incorporating earth tones into design, but suggest that the brighter colours found in nature are used more sparingly, and artificial vibrant colours are avoided. In nature, bright colours, especially when used with stripes, signify a warning such as poison or venom, but they are also used for seeking attention and attraction. Browning, Ryan and Clancy refer to colour under their "material connection with nature" principle and note that green may enhance creativity, and warmer colours are generally calming.[1]

Stephen Palmer and Karen Schloss (2010) proposed the ecological valence theory (EVT) of colour preference which ties in with evolutionary psychology and our innate response to the colour of natural foods and other "safe" colours. They state that "People are more likely to survive and reproduce successfully if they are attracted to objects whose colors 'look good' to them and avoid objects whose colors 'look bad' to them". They explain that people will be attracted to colours that elicit a positive reaction, such as blue and cyan reflecting clear sky and clean water, but repulsed by colours that generate a negative response, such as brown which can represent decay.

- *Light* – Kellert and Calabrese advise that natural light can be introduced into deep plan office using atria, clerestories, glass walls and reflecting colours and materials. I recall at a conference at the Danish Technical University, the delegates were all fascinated by the reflection of light on water surrounding the event venue. Browning et al. recommend dynamic and diffuse light that recreate the varying intensities of light and shadow that change over time in nature. Kellert and Calabrese note that "Beyond simple exposure, natural light can assume aesthetically appealing shapes and forms through the creative interplay of light and shadow, diffuse and variable light, and the integration of light with spatial properties". They also remind us that artificial light can be designed to mimic the spectral and dynamic qualities of natural light.
- *Water* – Kellert and Calabrese suggest that:

> The attraction to water can be especially pronounced when associated with the multiple senses of sight, sound, touch, taste, and movement … Water in the built environment is often most pleasing when perceived as clean, in motion, and experienced through multiple senses (although at muted sound levels).

Water features such as fountains and aquariums can be introduced indoors or, if possible, views of outdoor water bodies and wetlands provided. Browning and colleagues found that repeated experiences of water do not diminish interest over time, so one small water feature may be adequate. Experiments in using soundscapes for sound masking use water sounds in conjunction with a small water feature; otherwise, the water sound alone can be disorientating (Hongisto et al., 2017).

- *Variability* – In their 14 biophilic design principles, Browning, Ryan and Clancy recommend both non-rhythmic sensory stimuli and thermal and airflow variability. The former relates to continually experiencing varying stimuli in nature such as birds chirping and leaves rustling, whereas the built environment has a deliberate steady-state condition. In contrast, "space with good non-rhythmic sensory stimuli feels as if one is momentarily privy to something special, something fresh, interesting, stimulating and energizing. It is a brief but welcome distraction". The latter refers to how variability in temperature and ventilation feels refreshing, invigorating and comfortable, and how such spaces provide a feeling of both flexibility and a sense of control.

Browning et al. note that:

> conventional thermal design tries to achieve a narrow target area of temperature, humidity and airflow, while minimizing variability … laboratory-based predictive models assert that 80% of the occupants would be satisfied at any given time… An alternative approach is to provide combinations of ambient and surface temperatures, humidity and airflow, similar to those experienced outdoors, while also providing some form of personal control (e.g. manual, digital, or physical relocation) over those conditions.

Kellert and Calabrese suggest that "These conditions can be achieved through access to the outside by such simple means as operable windows, or by more complex technological and engineering strategies" and "Processed air can also simulate qualities of natural ventilation through variations in airflow, temperature, humidity and barometric pressure".

- *Prospect and Refuge* – Prospect refers to views of surrounding settings that allow people to perceive both opportunities and hazards; and in landscaping, prospect is usually characterised as the view from an elevated position across an expanse. This natural feature may be achieved by offering vistas to the outside, visual connections between interior spaces using balconies and open staircases (Figure 6.7), glass partitions and contiguous space.

*Figure 6.7* Prospect provided by interconnecting staircases (along with a moss wall).

Refuge relates to a sense of safety and retreat, achieved by providing spaces that allow withdrawal. Browning and colleagues note that "A good refuge space feels separate or unique from its surrounding environment; its spatial characteristics can feel contemplative, embracing and protective, without unnecessarily disengaging". Provide refuge by offering spaces that are reserved for reflection, rest and relaxation, spaces that have good visual and acoustic privacy, spaces that feel cosier using subdued lighting and lowered ceilings and spaces that are slightly detached from the main place of work (Figure 6.8). Such spaces can be indoors or, with good shelter, outdoors.

- *Multisensory* – Biophilic design solutions that bring nature into the workplace should operate at different levels and scales and be multisensory. All too often design focuses on the visual aesthetic but consider the non-visual connection with nature. As Browning, Ryan and Clancy (2014) suggest, explore the "Auditory, haptic, olfactory, or gustatory stimuli that engender a deliberate and positive reference to nature, living systems or natural processes". The use of soundscapes, like flowing water or birdsong, is a good example of an auditory connection to nature, and I previously mentioned introducing pleasant odours in the office.
- *Attachment* – Kellert and Calabrese mention the feeling evoked through the cultural and ecological connection to a place:

*Figure 6.8* Refuge provided by quiet booths with coloured moss walls at the Crown Estate (image courtesy of the Crown Estate).

Culturally relevant designs promote a connection to place and the sense that a setting has a distinct human identity. Ecological connections to place can similarly foster an emotional attachment to an area, particularly an awareness of local landscapes, indigenous flora and fauna, and characteristic meteorological conditions.

Use of natural and indigenous materials helps with the connection as well as being a nod to vernacular architecture. A timeline showing the history of an established company also creates a connection.

- *Patina of time* – Nature is always changing and in flux, and according to Kellert and Calabrese, "People respond positively to these dynamic forces and the associated patina of time, revealing nature's capacity to respond adaptively to ever changing conditions". Use of naturally aged and weathered materials provides a sense of the passage of time.

## Building standards

The international standards mentioned in the previous chapter include biophilic design elements such as daylight, natural ventilation and sound. However, the recommendations in standards are quite quantified and prescriptive, ignoring qualitative requirements, individual preferences and innate human needs such as variation.

Guidance on biophilia is included in the earlier version of *WELL* under two features. For the first feature "Biophilia I – Qualitative", points are awarded for providing a biophilia plan that describes (i) how nature is incorporated through environmental elements, lighting and space, plus (ii) how patterns in nature are included in the design, and (iii) shows that there are sufficient opportunities for human–nature interactions within and outside the building. For the "Biophilia II – Quantitative" feature, the design criteria are that (i) 25% of the site features either landscaped grounds or rooftop gardens accessible to building occupants and consists of, at minimum, 70% plantings including tree canopies, also (ii) plants cover at least 1% of floor area per floor and a plant wall per floor is provided and (iii) there is at least one water feature for every 9,290 m² floor space that is approximately 1.8 m tall and exposes the occupants to the sight and sounds of moving water.

Whilst the biophilic features and criteria mentioned in *WELL* are relatively basic, they at least represent a growing need for understanding biophilia and how to integrate it in workplace design. Undoubtedly, it will be some years before national standards and legislation fully recognise biophilic needs. So, consider this chapter, *WELL* and the guidance from Stephen Kellert, Bill Browning and their colleagues as a head start.

## Note

1  I am not a fan of "colour psychology", and caution is required when predicting the impact of colour on mood, behaviour and performance. Research into the effect of colour is inconsistent but does show that colour preference is affected by culture, age, gender, task and mood.

# 7   A plan comes together

As we are different "animals" with different individual preferences, the choice of a range of environments is key. One criticism of the modern interpretation of the open plan office is that there are few partitions and screens, resulting in rather daunting seas of desks. The work-settings in the *Landscaped Office* therefore vary in the level of screens, affording different amounts of privacy depending on whether increased interaction or reduced distraction is required. The design and structure of the furniture in the work-setting may consist of the following structures.

- *Room* – A fully enclosed workspace reaching the ceiling with a door for maximum confidentiality and minimum noise distraction. Room heights are typically 2.7 m, or for an exposed ceiling slab 3.6 m or more. Good acoustic design, providing sound containment and reducing noise into the room, is critical. Consider using glass partitions, with manifestations (transfers), to provide daylight ingress and a sense of connection whilst allowing some visual privacy. Temperature control and good air quality are also necessary in these enclosed spaces. Rooms are usually tailored designs, but demountable modular partitions may also be used. In a predominantly open plan environment, rooms provide surfaces (inside and outside) for displaying and sharing information, whether electronically using screens or through writable walls and whiteboards.
- *Pod* – A pod is also an enclosed space, but it is usually a self-contained furniture solution so free-standing and not part of the building infrastructure. As such, enclosed pods are typically 2.2–2.5 m in height. Pods may be cylindrical in shape or cubes/boxes. They usually have glass sides for transparency, but glass all round cylindrical pods are often disliked and referred to as "goldfish bowls". The pods may be exposed to and dependent on ambient environmental

DOI: 10.1201/9781003129974-10

conditions, which can degrade the acoustic properties, or they may be connected to the mechanical ventilation system. Pods usually facilitate 1:1 meetings, small groups or occasionally solo work. Pods, like booths, are usually furniture solutions and, as such, are more flexible and adaptable than rooms.

- *Booth* – Booths do not have a ceiling but usually have at least three sides that are high enough to offer some enclosure. The height of a booth is typically circa 1.6–1.8 m, so the occupant cannot be easily seen when sat down. A booth is usually constructed of sound absorbing material to minimise noise distraction and facilitate quiet work. They may have an open back, like a carrel, or have a more wrapped around enclosure.
- *Open area* – The more open plan space is usually used for desking. As well standard desks, fitted with keyboards and screens, touchdown spaces may also be provided for occasional use. The open plan may also include informal meeting areas, but if the meeting space is not dedicated to a close-knit team for their sole use, it will need screening or locating away from desks to reduce noise distraction to other nearby teams.

Work-settings of different height and construction break up the open space, aiding visual and acoustic privacy, and generate intrigue and interest by creating small pockets of space.

### 'Til desk do us part

Traditionally, the available work-settings in most offices were limited to open plan desks, cubicles, private offices and meeting rooms. Back in 1995, in their book *Understanding Offices* (my very first workplace book), Joanna Eley and Alexi Marmot refer to the open plan and large open plan workstation plus the enclosed office; they also mention ancillary spaces such as copy areas, kitchens/pantries and meeting rooms. Figure 7.1 shows a standard desk, larger desk and private office along with a study office and paired office. Previously, larger desks may also have been provided for managers, offering them additional surface area for paperwork and extra space to meet their team members. However, a reduction in hardcopy paperwork and larger flat-screen monitors makes the larger desk surface redundant. It is also preferable to meet staff in a meeting area rather than in the middle of the open plan. So, larger "manager" desks are usually not included as part of the settings for the *Landscaped Office*. Nevertheless, the desk should be sufficiently large to respect personal distances and help alleviate noise. Sit-stand desks, mandatory in most countries on the

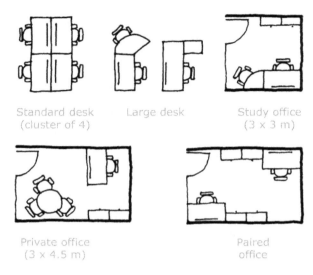

Standard desk          Large desk          Study office
(cluster of 4)                             (3 x 3 m)

Private office                                    Paired
(3 x 4.5 m)                                       office

*Figure 7.1* Traditional standard desks and private offices.

European mainland, offer staff a break from sitting, which is better for our health (Buckley et al., 2014).

Study offices are increasingly becoming more common; they are basically a more compact private office, used either when space is limited or in more traditional and hierarchical workplaces sometimes used to indicate a less senior position than a full-sized office. A well-designed study office offers privacy control, confidentiality and is usually as functional and practical as the larger ones. Paired/shared offices are also a means of saving space and are used more in academia and law firms. They work better for academics when, due to their teaching duties, etc., both occupants are unlikely to be in at the same time. They are more suitable for lawyers who work closely together, usually a partner and junior team.

In the more egalitarian *Landscaped Office*, private and paired offices may be required in extenuating circumstances due to role, personality, sensory sensitivity, inclusivity, culture, etc. However, they are usually not included because there are other options available for facilitating concentration, confidentiality, collaboration and personal factors.

If space efficiency and cost of construction, such as the supporting beams and legs, do not constrain the desk layout then alternative desk configurations can be explored. Figure 7.2 illustrates some alternatives to rectangular desks, such as a vertical (or sideways) offset which means

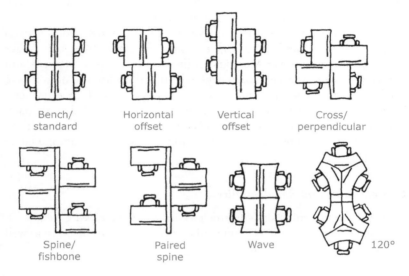

| Bench/ standard | Horizontal offset | Vertical offset | Cross/ perpendicular |

| Spine/ fishbone | Paired spine | Wave | 120° |

*Figure 7.2* Alternative desk configurations.

that the occupants no longer directly face each other, thus helping reduce speech interference, especially where there is much telephone use. The shape of the desk may also be used to create different configurations, such as 120° desking which forms clusters of three desks or six when clusters are paired and sometimes quaintly called a "dog bone". Alternative desk configurations add interest and can help reduce visual and acoustic distractions.

## A menagerie of work-settings

The nature of work is continually changing with, for example, a shift from processing to creation. Furthermore, workstyles are also continually changing, with increasingly more people working away from the office and more likely to conduct process and focussed work at home. So, the balance of desks to alternative work-settings in the office is also shifting, with fewer desks and more alternatives provided. Nowadays, thanks to advances in design and furniture, a whole menu of work-settings can be made available to office inhabitants rather than providing limited choice and homogeneity. Office occupants require a range of work-settings that match their requirements and preference. These alternative (non-desk) work-settings depend on the occupants' primary work activities, their personality, their sensory sensitivity, their biophilic needs, their mood, etc.

The following is a non-exhaustive list of alternative work-settings proposed in the *Landscaped Office*. Some are taken from the guides produced by Juriaan van Meel and colleagues (van Meel, Martens and van Ree, 2010; van Meel, 2020), whereas the others are the ones I have used in my own workplace projects or seen used successfully elsewhere. The number of alternative work-settings required is established through consultation with the anticipated office inhabitants backed up with observation studies. The size of each setting is based on design guidance and experience.

### Concentration and confidentiality

As per my overarching design principles for the *Landscaped Office*, work requiring concentration and confidentiality is required in the office as well as collaboration and creativity.

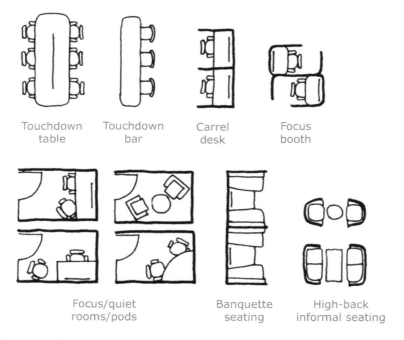

Touchdown table    Touchdown bar    Carrel desk    Focus booth

Focus/quiet rooms/pods    Banquette seating    High-back informal seating

*Figure 7.3* Alternative work-settings.

- *Quiet zone* – A designated quiet area for working without distraction. The zone may include carrel desks, with high screens, as well as focus booths and pods. The design may mimic a library or reading room offering a visual clue to the expected behaviour.
- *Carrel* – A standard desk but with high screens on three sides used for focussed and quiet work. Carrels are usually provided in designated quiet zones.
- *Focus/quiet space* – This could be a booth, pod or room facilitating solo focussed work, with minimised distraction (Figure 7.3). As well as quiet work, the space may be used for confidential calls or video conferencing, thus reducing the distraction for nearby colleagues. These spaces usually include power, connectivity, monitors and keyboards. Prior to wholesale videoconferencing, the rooms and pods will have included a spider or star phone for teleconference calls.
- *1:1 space* – A pod or room for facilitating intimate and private one-to-one meetings. The room may also double as a focus space. Different furniture layouts in the 1:1 rooms create a different atmosphere and indication of the activities that will take place in the space.
- *Phone booths* – In open plan areas, nearby single-person phone booths facilitate personal or highly confidential calls. Focus pods and booths can also be used for such activity.

### Collaboration, creativity and connectivity

The office is increasingly being used to foster creativity and innovation as well as to facilitate collaboration and connecting people.

- *Touchdown space* – Typically, a large desk or table, sometimes referred to as a "kitchen table", provided for intermittent laptop work or to facilitate visitors or groups of cross-functional colleagues occasionally working together. Touchdown bars with stools are often provided but need to be well located with some privacy if they are to be regularly used. Access to power and internet connectivity (usually Wi-Fi) is required, but touchdown is aimed at laptop use and does not usually include keyboards or monitors like standard desks.
- *Banquette seating* – Usually, a rectangular or circular booth used to create an informal semi-private meeting space. A rectangular booth may have two to six seats, but the two-seaters are popular for solo work. Such banquette seating may also be covered, providing a sense of refuge and more privacy, and when adjacent and in line are sometimes referred to as a "railway carriage". Integrated screens facilitate knowledge sharing. Circular banquette seating may be larger and sometimes referred to as huddle spaces.

- *Meeting space* – Meeting rooms vary in size typically accommodating from 2 to 12 people in the main office area along with more central larger meeting rooms for 12–36 people in a conference or meeting suite. Those rooms may be even larger, depending on requirements such as training and events and are expandable using folding partitions. Meeting rooms are primarily used for formal meetings and presentations and so require good acoustic and visual privacy. Meeting rooms should also have easily accessible power and connectivity plus well-designed audio-visual technology facilitating face-to-face and hybrid physical/virtual meetings. Pods may be used for smaller private meetings in the main office space. Meeting rooms and pods do not necessarily need to be rectangular; Siren Design in Australasia developed pentagonal-shaped rooms and Peter Andrew and colleagues at CBRE created

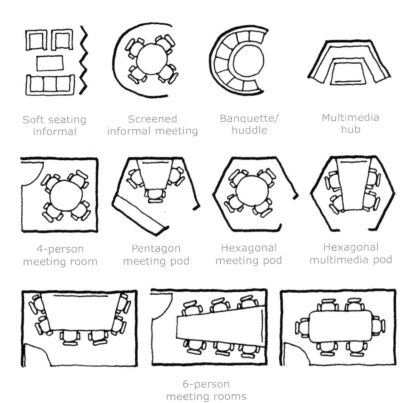

Soft seating informal    Screened informal meeting    Banquette/ huddle    Multimedia hub

4-person meeting room    Pentagon meeting pod    Hexagonal meeting pod    Hexagonal multimedia pod

6-person meeting rooms

*Figure 7.4* Meeting spaces.

the stretch meeting room (a stretched pentagon). Hexagons tessellate well and are also an ideal shape for a meeting or focus pod, creating a honeycomb of spaces (Figure 7.4). Meeting rooms may also be designed to facilitate stand-up meetings, which offer a break from sitting and are often shorter. The outside indented partitions of pentagon and hexagonal pods also facilitate stand-up informal meetings.

- *Project room* – This is a cross between a meeting room and brainstorm area that may be allocated to a team for the duration of their project. The rooms allow the team to work closely together and to share and display confidential material, without having to pack it away each evening. Hanging or reversible boards allow more than one team to use the space at different times.
- *Brainstorm area* – Usually, a meeting room with additional whiteboards, whitewalls or smart boards (electronic whiteboard). However, such spaces do not need to be fully enclosed if located slightly away from the main desks. The furniture and decoration may be slightly different, it may be more alternative, funky and colourful, to signify that it is not just another standard meeting room.
- *Informal meeting areas* – A small table and set of chairs in the open plan support impromptu informal meetings. However, if located close to desk areas, then, some screening will be required to reduce noise distraction; the exception is for close-knit teams whose members may benefit from listening in on their colleagues' meetings. High-back seating offers some acoustic privacy and may consist of two single wing-back seats opposite each other or two-person seating. Informal meeting areas may comprise soft seating and sofas for a more informal setting. Some organisations are installing amphitheatre-type seating for impromptu gatherings. As with meeting rooms, consider facilitating informal stand-up meetings.
- *Multimedia hub* – These are usually furniture solutions, such as Steelcase's "media:scape", that facilitate collaboration in the open plan. They often include some low screening, informal seating and the capability to connect to an integrated shared screen.

### Contemplation and care

The need for contemplation, relaxation and social interaction should not be underestimated in the workplace.

- *Breakout space* – An area allowing the occupants to break away from the main desk area. Breakout spaces may facilitate informal meetings and social interaction. Breakout space is usually located near a small kitchen or vending area and includes a range of furniture styles. Consider supplying healthy food options, especially, in those chocolate-laden vending machines. Larger breakout spaces may include recreation facilities such as pool, table tennis or video games.

- *Chillout area* – An alternate breakout space providing a setting for contemplation, relaxing or reading alone. Like the quiet zone, it may mimic a reading room or library with a calming environment with subdued lighting, materials and soft furnishings. Such areas often include planting and, occasionally, an aquarium. Sleep pods and nap areas have also been introduced by some organisations.

### Core and common spaces

Of course, other work-settings will be required in the *Landscaped Office* such as a kitchen/pantry/vend area, filing and storage, print and copy, cleaning cupboards, and comms cabinets plus other specialist spaces. The office building will also require other central spaces possibly including a reception and waiting area, staff restaurant and kitchen, gym/welfare centre, conference suite, concierge, security, etc., and buildings also have core areas including the stairwells, elevators, ducts, mechanical plant, service areas, goods-in and waste, etc. The appointed interior architect will design and bring all the space components together, ensure that they work well collectively and are supported by the building services and infrastructure.

## Planning the landscape

There is good, and there is bad office design; amongst a host of factors, the latter is mostly due to increased density and lack of partitions creating large open areas rather than discrete zones. The following fundamental space planning principles underpin the success of the *Landscaped Office*. As it is a revised version of a long-standing concept, much of the following is existing best practice already seen in many modern offices. Nevertheless, such principles are often forgotten or ignored, not applied in their entirety, and some are new.

### Workplace layout

Plan the desks and other work-settings for comfort, not just efficiency, by incorporating the following recommendations.

- Adhere to local legislative and best practice space planning principles, including complying with all disability legislation, especially for wheelchair users.
- Typically provide 1,500–2,000 mm primary circulation routes and 1,200+ mm secondary circulation between desks; wider circulation space helps with acoustic privacy and a sense of space, that is, reduces feelings of overcrowding.
- It is a standard practice to place desks near to and perpendicular to windows to offer daylight but minimise glare. However, consider different (non-orthogonal) orientations of desks to add interest and create a more natural layout (but check glare is avoided); such planning also creates pockets of space for other work-settings such as collaboration areas.
- Arrange desks so no one has their backs to the primary circulation routes; different orientation of desks adds interest but avoid people being easily overlooked from behind or distracted by peripheral movement.
- Plan for 2.5 m or more between the edges of back-to-back desks but provide an absolute minimum of 2 m between desks (i.e. a 3.6-m pitch centre to centre).
- Partially break up the space every three or so desk clusters, or for each team (typically 8–12 staff), using pods, filing, high-back seating, planting, open bookshelves, mobile whiteboards, etc.
- Ideally, use clusters of two or four desks and a maximum cluster of six desks and avoid long runs of three or more desks in a row.
- Place any air vents over circulation routes, rather than above desk chairs, to prevent draughts and consider the placement of overly bright light sources.

### Workspace zones

Discrete zones help with visual and acoustic privacy as well as create a sense of belonging and interest.

- A contiguous single space is often preferred by managers, but continuous rows of desks should be avoided by introducing other work-settings and semi-partitions to create a more landscaped environment with distinct zones along with nooks and crannies for interest and refuge.
- Desks should ideally be arranged in distinct zones, for use by teams; zones do not need to be segregated but may be defined by some semi-partitioning or different work-settings such as pods or screened off informal meeting or filing (Figure 7.5).

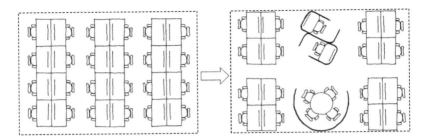

*Figure 7.5* Break rows of desks into zones using alternative work-settings.

- Zone the open plan into areas that better support specific activities, providing the right proportion of work-settings that support those activities, then empower and encourage staff to use those settings as and when required.
- Consider zoning the open plan into areas that are cooler, warmer, quieter, buzzier, lighter, dimmer, etc., especially if agile working, with unassigned seating, is introduced.
- Contain noise away from the open plan desks using enclosed and semi-enclosed nearby meeting spaces.
- Alternative work-settings are also required nearby for focussed or confidential work.
- Room partitions should be glazed to allow maximum daylight, but manifestations may be required for visual privacy.
- Use easily accessible staircases, ramps and bridges to connect floors and encourage movement. Creating novel, non-direct, routes to the desk may also aid creativity.
- Add further vertical variety using slightly raised work areas, with steps and ramps, and amphitheatre-style social and informal meeting areas.

### Work-settings

Alternative work-settings help create a different look and feel as well as a more humanised workspace.

- Consider alternative layouts to the rectangular desk or provide non-rectangular desks which provide a more organic layout, appealing to our innate biophilic needs and increasing the sense of space. However, they can take up more space and may be more costly than the usual blocks of rectangular bench desks (Figure 7.6).

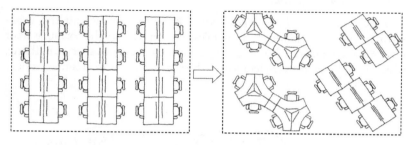

*Figure 7.6* Non-rectangular and non-orthogonal desking has a more organic feel.

- Provide a sufficient desk size to reflect personal space and help reduce noise; I prefer 1,800 × 800 mm desks (similar to the average male arm span, i.e., reach) which allows sufficient personal space and helps reduce speech interference, and my recommended minimum is 1,600 × 800 mm (similar to the average female arm span) which still fits two monitors if required (and is also in-line with the WHO's 1.5 m social distancing rule in response to the Covid-19 pandemic as there is 1,600 mm to the side, in front and behind the person sat at the desk).
- Desk screen heights should be kept low, for clear lines of sight, but high enough to reduce noise distraction from speech, for example, 1,400 ± 100 mm.
- Consider flexible/moveable (with wheels) and stackable/foldable furniture in meeting spaces so that it can be easily reconfigured providing multifunctionality.
- Where practical use furniture solutions, such as pods and booths, to provide a more flexible and adaptable environment, the occupant's activities and requirements may change over time, and indeed, the occupants may change.
- Use low (three-high) storage/filing units throughout the open plan space (if required) with any taller cabinets placed against walls.
- Avoid taking up valuable floor space with storage; storage requirements are ascertained through consultation and storage audits, but with more agile working and advances in electronic storage and document management, less storage is required, other than personal lockers/storage for all staff.
- Storage units between clusters of desks and between the end of a row of desks and the primary circulation route provide some acoustic and visual privacy.

## Bringing it all together

The essence of the *Landscaped Office*, bringing together all the work-settings and space planning principles, is captured in the example floor plate shown in Figure 7.7. The space plan is not related to a specific building, which would require consideration of the core areas, escape routes (primary

*Figure 7.7* Example of the Landscaped Office, for illustration purposes only.

circulation), daylight ingress or orientation, etc. Neither does it relate to a specific organisation; the starting point of any design and space plan is understanding the business vision and needs of the organisation, the core work activities and individual requirements including personal, psychological and physiological factors.

Rather, the example space plan is a showcase for the *Landscaped Office* illustrating how it is more structured than *Bürolandschaft* but not as orthogonal as typical open plan workspaces. *Bürolandschaft* plans often show random single desks, whereas the *Landscaped Office* offers small clusters of two to four desks, six maximum or single booths and pods and an array of distinct smaller team or activity zones. The space plan illustrates a quiet zone (top right), other notional zones and a central breakout/social space along with a variety of other spaces and furniture arrangements.

### Look and feel

The look and feel, that is the aesthetic and ambience, of the space is also important, and it is recommended that the design better reflects the requirements of its inhabitants, derived from environmental psychology, evolutionary psychology and other disciplines already discussed in the early chapters.

Biophilic design principles are fundamental to the *Landscaped Office* and, where possible, the furniture and finishes should be natural materials, and the patterns and colours used should reflect those found in nature. In addition, well-managed planting should also be provided throughout the floorplate, as plants assist with acoustic and visual privacy. The example space plan, shown in Figure 7.7, shows plenty of plants dotted throughout the space and in the central interaction zone, but also considers views out onto nature. Likewise, provide indoor–outdoor-type spaces or outdoor terraces for breakout areas, which are good for daylight as well as accessing nature. Well-designed lighting, glazing and water features can all be utilised to satisfy our innate biophilic preferences.

As in any predominantly open plan environment, acoustics is key; this is partly resolved by reducing the density and breaking up the space using alternative work-settings, screen, open bookshelves and plants, but also consider the ceiling, floor and wall finishes and how to add absorption. Colour, artificial lighting and natural patterns can be used to create different moods such as calming versus stimulating spaces. Visual clues signify the anticipated behaviours in the space, for example, books and bookshelves indicate quiet, whereas coffee points, round meeting tables and games (pool, table tennis) suggest social interaction. As well as the look and feel of the *Landscaped Office* reflecting human needs so must the indoor environmental conditions, discussed in Chapter 9.

### Reckoning occupational density

Occupational density, the Net Internal Area (NIA) divided by the number of open plan desks and private offices, is one indicator of space efficiency for traditional offices. As mentioned, the CRE industry, particularly in the UK, has historically focussed on increasing density to save property costs. However, calculating the density is not so clear cut for the *Landscaped Office* with alternative work-settings. Consider the example space plan, shown in Figure 7.7, which is approximately 900 m$^2$ with 68 standard desks and carrels equivalent to circa 13 m$^2$ per desk. This may be considered generous by those in the CRE industry who are pursuing high densities in order to reduce the property costs. Nevertheless, the additional 15 pods and focus rooms provide a total of 83 desks, which is equivalent to 11 m$^2$ per desk and quite space efficient, but not as ridiculous as the higher floor densities of 8 m$^2$ per desk often expected.[1]

One problem with using density as the sole metric of space efficiency is that it does not provide an indication of the level of support space. Higher densities can be, and often are, achieved by reducing the support space and cramming offices with desks. The proportion of support space must therefore also be considered along with the density, and best practice is typically between one-third and one-half of the space. The primary circulation space also needs to be factored in, too much may be space inefficient, but too little may impact on egress and limit mingling and interaction outside of the desk areas.

On the example space plan, the standard desks and associated secondary circulation cover approximately half (50%) of the space, but one-third (35%) of the space is used to provide 36 meeting spaces, for two to six, plus breakout, with the remaining 15% providing the legally required primary circulation. Thus, the example plan is well-balanced, and if anything, in the *Landscaped Office*, the desk space should be reduced to around 40% to allow more alternative work-settings.

Saving space and reducing property costs should not be the key driver for a workplace strategy and corresponding design and layout. However, if there is limited space available to an organisation, then agile working offers a better workplace solution. Agile working is discussed in detail in the next chapter, but one key component is unassigned seating with desk sharing. Implementing desk sharing can either help release space or alternatively save space by accommodating more staff, termed spaceless growth. If desk sharing is applied to the example space plan (shown in Figure 7.7) at a ratio of 10:6 staff per desk, then the layout could accommodate circa 110 people, equivalent to approximately 8 m$^2$ per person which is highly efficient and in-line with the typical target floor densities.

# Note

1 Occupational densities refer to the whole building NIA, including the main office floors plus conference suites, staff restaurant and reception. This is a different metric to on-floor densities. In the UK, the BCO (2014) indicates that an "overall occupation on floor based on 1 person per 8 $m^2$ and their density study" (Bedford et al., 2013) found that "workplaces on a purely work floor might be planned to a high density (say, 8 $m^2$)".

# 8 Choice, sharing and agile working

Back in Chapter 4, I chartered the evolution of office design, but a relatively recent workplace trend is for agile working in which the space is a balance of desks and other activity-based settings designed to facilitate specific work activities such as collaboration, creativity, concentration and confidentiality. Agile working with its unassigned desking (hot-desking/hoteling) and remote/home working, alongside the choice of work-settings, complements the *Landscaped Office*. This combined Agile Landscaped Office (ALO) creates a workplace that caters for individual requirements, thus enhancing the wellbeing and performance.

Agile or activity-based working (ABW) has been implemented in various forms since the mid-1990s, with origins back in the 1970s. Juriaan van Meel (2011) postulates that these workplace strategies are not new, and "the concepts of mobile offices … and flexible workplaces all originate from the end of the 1960s and the early 1970s" in organisations such as IBM. Apparently, American architect Robert Luchetti first mentioned "activity settings based environments" and "multiple settings to support the variety of performance modes" in the late 1970s (Stone and Luchetti, 1985). However, it was Dutch insurance firm Interpolis that introduced the first successful version of agile working, which they called "flexible working" in 1995, but they were pipped at the post by US-based advertising agency Chiat/Day with their famously ill-fated attempt in 1994.[1] At a similar time, Frank Duffy, one of the founders at workplace consulting practice DEGW, introduced the *Hive, Den, Cell* and *Club* office typologies (Duffy, 1997). The *Club* is the most autonomous and interactive working environment and the one with unassigned seating; Duffy described it as an office where "individuals and teams occupy space on an as-needed basis, moving around it to take advantage of a wide range of facilities".

DOI: 10.1201/9781003129974-11

## Types of agile working environments

Agile working may have had various guises over the years, each version with a slightly different focus, but they all include three core components, illustrated in Figure 8.1 and listed below.

- *Flexible work patterns* – At the heart of agile working is flexibility in terms of time as well as space. Choice and autonomy over working hours and times in and out of the office are one of the benefits for staff. Staff may be offered full flexible working hours where they control the start and end times of their working day or they may be offered other options such as part-time work, job share or compressed hours. Secondments and career breaks are also flexible working options. Since the advent of Covid-19, some organisations have taken a more structured approach to the times that staff are in and out of the office, that is, assigned office days rather than full flexibility. Flexible working practices and policies tend to be managed by the organisation's HR team.
- *Remote working* – Another key component of agile working is the ability to carry out work away from the desk. Remote working primarily refers to working outside the office such as working from home (WFH),[2] elsewhere or travelling on business. The latter are sometimes referred

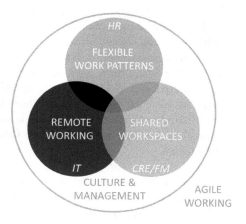

*Figure 8.1* Key components of agile working.

to as the "third place", with the home being the "first place" and the office the "second place" where people spend most of their time. Third places include cafés, pubs, libraries, co-working spaces, business clubs, colleague's homes, client's offices, etc. Remote working also refers to working in the office but away from the desk such as in meeting spaces, focus pods or breakout areas. Telework and telecommuting are other common terms for remote working, but perhaps with more emphasis on not having to commute to an office. Not being tethered to a desk is dependent on mobile technology (e.g. laptops, phones and handheld devices), good (wireless) connectivity and the ability to access the organisation's data and applications, etc., from outside of the office. Remote working is therefore predominantly reliant upon the services of the IT team. However, it also requires a change in mindset, culture and management.

- *Shared workspaces* – The final component is that of shared workspaces. Agile working environments include a wider variety of work-settings for shared use by all occupants. For example, rather than just rows of desks and meeting rooms, the space might include informal meeting areas, collaboration spaces, focus pods and booths, 1:1 rooms, telephone booths, quiet zones with carrels, breakout and spaces for social interaction. In addition to sharing the work-settings, the occupants may also be required to share desks, through unassigned desking or hot-desking.[3] In highly mobile organisations, staff may be expected to find a free workstation anywhere in the building, termed free address and disliked by most, or they may book a workstation in advance, termed hoteling. However, for most organisations, the staff congregate in desk areas with their colleagues termed a team or home zone or neighbourhood. As well as standard desks, there may be touchdown desks to use for short periods of time. Agile working environments are usually open plan workspaces and do not have assigned private offices, but some organisations have experimented with "hot-offices" or shared study/workrooms. Shared workspaces require design and management so are usually curated by the Facilities Management (FM) or Corporate Real Estate (CRE) teams.

For an agile workplace to be truly successful, it requires all three of the above components to be implemented and working together. Experience shows that agile working strategies implementing just one element fail and those with two are more difficult to sustain over time. Underpinning the three components is choice – the ability for employees to choose where and when they work. Offering choice is dependent upon the organisational culture and on management trusting and empowering the workforce to work when and where they are most efficient and productive whilst maintaining work-life balance and wellbeing.

The various terms for agile working relate to their originators and emphasis. The following list describes the most common terms and strategies.

- *Agile working* – Agile working is the most recent phrase adopted by the workplace industry, particularly in the UK, and is used throughout this book as an umbrella term. The phrase agile working is more business focused than the other terms, emphasising the way in which work is carried out and businesses are managed. The Agile Organisation defines it as "bringing people, processes, connectivity and technology, time and place together to find the most appropriate and effective way of working to carry out a particular task" (Allsopp, 2010). Agile working is derived from "business agility" related to the speed of responsiveness of the organisation to adapt its workflow and workforce to changing customer/market demands. Confusingly, the term "agile working" also refers to "agile project management" with its scrums and sprints, popular in the technology sector. Whilst definitions of agile working do not implicitly refer to unassigned seating, like other variations, it tends to be par for the course when implemented as part of a broader workplace strategy.
- *Activity-based working* – The primary emphasis of ABW is on providing a range of work-settings that support the multitude of different activities that take place in the office rather than depend on all activities taking place at a desk or in meeting rooms. Veldhoen + Company (2018) propose that "the ABW approach recognises that people perform different activities in their day-to-day work, and therefore need a variety of work-settings supported by the right technology and culture to carry out these activities effectively". Some definitions of ABW infer that unassigned workstations are not required, but most implemented versions include hot-desking. ABW also refers to remote working such that the work-settings extend to those away from the office as well as inside it; Veldhoen + Company state that "ABW creates a space that is specifically designed to meet the physical and virtual needs of individuals and teams". Erik Veldhoen is credited with coining the phrase ABW in his 2004 book *The Art of Working*, but Veldhoen + Company actually implemented the ABW philosophy at Interpolis a decade earlier. Veldhoen + Company also stress "ABW's heavy emphasis on the creation of a culture of connection, inspiration, accountability and trust empowers individuals, teams and the organisation to perform to their potential".
- *Flexible working* – This is an older term for agile working with emphasis on flexible working hours. The term was popular with UK public

authorities that offered flexitime and other flexible working arrangements. However, the term was also associated with occasionally WFH and better work-life balance by being able to choose when to come into the office. The term "flexible working" is seeing a resurgence and is preferred to "agile working" by some organisations to prevent confusion with agile project management.

- *Smart working* – The HR professional body CIPD (2014) define smart working as "an approach to organising work that aims to drive greater efficiency and effectiveness in achieving job outcomes through a combination of flexibility, autonomy and collaboration, in parallel with optimising tools and working environments for employees". It perhaps places more emphasis on individual performance and organisational productivity but also acknowledges using space and technology to support a different way of working.

- *New ways of working (NWW)* – A less-used phrase nowadays, perhaps because NWW has been around in some form since the 1970s. The origin of the term NWW is elusive, but HRzone (2017) describes it as "an initiative looking to boost flexibility and retention, largely by removing many of the barriers and management styles of the past and bringing them into line with a modern multigenerational workforce". Bill Gates (2005) introduced the phrase "New World of Work" to describe using technology to better facilitate collaboration between dispersed workers, and some organisations have used similar phrases to refer to the post-pandemic workplace rather than the more commonly used "new normal".

- *Remote working* – A broad-brush term defined as "the practice of an employee working at their home, or in some other place that is not an organisation's usual place of business" and "a situation in which an employee works mainly from home and communicates with the company by email and telephone" (Cambridge University Press, 2021). The emphasis is often on WFH and is also referred to as telecommuting or teleworking.

- *Hybrid working* – Hybrid, or blended, working is a more recent phrase borne out of the Covid-19 pandemic. It reflects simultaneous physical and virtual working, with a mixture of the staff working in the office or WFH. In a nutshell, hybrid working is agile working with more emphasis on the technology, and management, to connect the workforce and ensure that those not physically present nevertheless feel connected. Hybrid working is undoubtedly facilitated through seamless technology with videoconferencing platforms and collaboration tools, such as Zoom and Teams. However, it also requires additional management skills to ensure that all staff feel part of the team and to ensure the right balance of staff are in the office at any one time.

The common theme of the above definitions is offering choice of when and where to work and is reminiscent of the "anytime, anyplace, anywhere" Martini adverts.[4] In terms of the agile workplace, they imply an office base and access to other working environments rather than an organisation that is fully virtual. My own simpler definition is:

> Agile working environments offer people the choice of how, when and where to carry out their work activities to enhance their wellbeing and performance. It is supported by a well-designed office, with multiple work-settings, along with the option to work outside of the office. Agile working environments are underpinned by enabling technology, an appropriate organisational culture and aligned management style.

Due to the ambiguity of the overlapping agile working definitions, many organisations develop their own terminology and define their tailored version of agile working. For example, BP's *BlueChalk*, BT's *Workstyle2000*, Centrica's *Work:Wise*, Deutsche Bank's *DB New Workplace*, HSBC's *OpenWork*, IBM's *e-Work*, MEC's *Smarter Working*, RBS *Choice*, the UK Cabinet Office's *The Way We Work* and Unilever's *Vitality*.

## Benefits and barriers to agile working

### Advantages of agile working

In well-designed zoo enclosures, the animals choose when to be on show or when to hide in the shadows. Larger enclosures allow the animals to find their favourite spot or the area that best suits their activities. Agile working fundamentally offers the choice of when and where to work by providing a range of work-settings. This free-range environment better suits most workplace inhabitants than the limited offering of the typical *Workplace Zoo*. Office workers are complex, diverse, inquisitive and social animals who have a range of preferred working environments and enjoy some variety.

Case studies and feedback surveys, such as the *Leesman Index*, have shown clear positive benefits of agile working, and those where the staff have more empowerment over where and when they work are most successful. Leesman (Rothe, 2017b) compared agile working environments (actually ABW) with more traditional workplaces, and they also compared the more mobile and transient workers with the more static and fixed ones. Leesman reported that "the data consistently supported industry claims that ABW increased staff collaboration, productivity, pride and effectiveness" with the effect more marked for the more mobile workers. The Global Workplace Analytics (2020) analysis of 2018 American Community Service data conservatively estimates that a typical employer can save an average of $11,000 per remote worker, a "result of increased productivity, lower real estate costs, reduced absenteeism and turnover, and better disaster preparedness".

Advocates of agile working have diligently shared the benefits. For example, Oseland and Webber (2012) compiled 32 case studies demonstrating the benefits, categorised as shown in Figure 8.2 and listed below.

- *Performance* – Better business performance through reduced absenteeism, extended business hours, improved individual performance, increased cross-selling, additional fees, quicker to market, faster decision-making, enhanced team-working, improved customer satisfaction, increased profit, etc.
- *Personal* – Personal benefits such as reduced travel time and cost, more empowerment over work, improved work–life balance, lower stress, easier access to the leadership team, better quality office environment, choice of work-settings, more social and wellbeing space and increased comfort (in the office when it is zoned with different environmental conditions or when WFH).
- *Sustainability* – Reduced organisational and personal carbon footprint and reduced business and personal travel. In most cases, a smaller office

*Figure 8.2* Benefits of agile working.

footprint will be required, meaning less heating/cooling, lighting, etc., of empty space and less embodied energy for new workplaces.

- *Continuity* – Improved business resilience and continuity due to less disruption from poor weather, security issues, viruses and travel problems, resulting in fewer lost workdays.
- *Enticement* – Reduced staff attrition and increased staff attraction resulting in reduced training and recruitment costs. More flexibility over work patterns and time in the office is attractive to those with disabilities or young families. It also supports recruiting from a wider, more geographically dispersed, labour pool and flexibility may be more attractive to younger employees.
- *Efficiency* – Space savings and reduced associated property costs, plus reduced churn and associated costs. Implementing unassigned desking often reduces the amount of space require to accommodate the current workforce or enable more staff to be accommodated in the same amount of space, termed spaceless growth. Efficiency should not be the primary reason for introducing agile working, and if it is the focus, then it will most likely be less successful.

### Challenges of agile working

Of course, agile working is not without its challenges and is not always as successful as expected. As mentioned earlier, an agile working environment requires a combination of remote working technology, flexible working patterns and shared workspaces. It also requires the right culture and management and is dependent upon the organisation fully trusting and empowering the workforce to work when and where they are most efficient and can perform to their maximum capability.

A potential problem with agile working is a lack of sense of belonging to the organisation. My own research revealed that workplace loneliness is higher for home workers and other research found that workplace loneliness leads to lack of loyalty, a higher chance of leaving, poor colleague engagement and teamworking, and ultimately, loss of performance. Some people who WFH will feel much more disconnected than others. A key challenge for managers is to ensure that their team members are fully engaged and connected. In contrast, agile working is associated with trust, more empowerment and choice, all offering psychological benefits.

Agile working environments with hot-desking require a clear desk policy to operate, so that the desk is ready for use by the next occupant. Some staff are more reliant on personalisation of their desk than others – usually, those spending more time in the office, for example, 9–5 for 5 days per week rather than more mobile workers. Eric Sundstrom (1986), a prominent environmental psychologist, proposed that personalisation is a demonstration to co-workers that the workspace is in the occupant's zone of control, so a form of territorial behaviour. Van der Voordt and van Meel (2002) suggest that the reasons for personalisation are practical, marking territory, creating recognition and expression of identity. Scheiberg (1990) suggests that personalisation is used as an unconscious outlet of emotions, and regulation of emotions in the workplace is key for wellbeing. Malmberg (1980) suggests that territorial behaviour is evolutionary as it is a common behaviour shown by all primates.

Therefore, another challenge of agile working is providing an alternative means of personalisation. This may be achieved by allowing personalisation of laptops, lockers and caddies. Caddies are the containers for personal items like stationery stored in lockers and used at the desk; I was impressed with one organisation who provided art materials and ran an afternoon session with their staff on how to decorate their caddies. In addition, switch the personalisation to the team zone by enhancing the team identity, having shared exhibits of memorabilia and displays of work projects and social events. As more people adopt WFH for a few days per week, the centre of gravity will shift from the office, and the expectations of the office will change so perhaps the need for personalising a desk will be less essential. The new working environment also needs to allow for those with special furniture and equipment, due to occupational health needs.

In pre-pandemic agile working environments, the staff may have used several desks during the day between meetings, etc. Going forward in the post-pandemic era, hot-desks are likely to be used by one member of staff each day rather than shared throughout the day as they become vacated. Therefore, they will require more rigorous cleaning, perhaps overnight by cleaning staff and just before use by the occupants, using alcohol wipes, etc.

### Work Patterns

The amount of time spent working in and out of the office is dependent upon role, personality, management style, culture, personal circumstances and other factors. At the heart of agile working is choice and having the flexibility to work in and out of the office that supports work-life balance as well as performance. Some organisations, especially those new to adopting hybrid working, prefer to implement a rota of when staff are in and out, either on a daily or weekly basis, but this smacks of forced choice and inflexible working so is not advised. Nevertheless, it is good practice for team members to meet up in person once per week, and this may require agreeing on which days teams are in and out of the office. The most favoured day for WFH is Friday which will need managing if the office is not to be left empty; I used to hold my team meeting on Friday as we could easily find meeting space, it sets us up for the forthcoming week and we would adjourn to the pub after work. Of course, some staff work on multidisciplinary or cross-team projects, so they will need to coordinate their diaries with project colleagues.

As well as coming into the office on different days of the week, the work hours themselves may vary daily. For example, people who commute to the office may prefer to miss the rush hour. It's bizarre that in an age when most large businesses have offices around the globe or offer a 24–7 service that many people still have standard office hours, which results in the morning and evening rush hours. Additionally, employees with a young family may wish to work around school start and end times, etc. Flexible working hours also allow staff to accommodate home deliveries or personal appointments occurring during typical working hours. The flexibility allows the staff to complete their contracted work hours without being pinned to specific times.

It also makes sense to work at the time of day when we are most effective and take a break when we are not. Some people are naturally "morning larks" and others are "night owls", that is, people are different chronotypes. In one of my on-line surveys, I asked the participants what time of the day they are most productive. Approximately half of my sample report being most productive in the morning and half in the evening. One-third (32%) said that they are most productive early morning before 9 am, with a further 13% most productive before noon. At the other end of the scale, one-quarter (24%) consider themselves more productive late evening after 8.30 pm, with a further 22% more productive early evening 5.30–8.30 pm. Other research (e.g. Jaffe, 2015) found that night owls outperformed morning larks on most intelligence measures, in particular, working memory and processing speed. In contrast, other research found that morning people tended to be more agreeable, conscientious, proactive and procrastinate less.

Two problems with flexible working hours are (i) staff work at different times so may miss out on contacting colleagues with a different work pattern and (ii) staff may feel that they need to be available 24/7. It's not unheard of that some people work unusual or extended hours, late into the night and at weekends. First, such work patterns may not be productive, and may be indicative of other work-related or personal issues and, so are probably not good for long-term wellbeing, and second it is certainly not acceptable for said people to expect their colleagues to work similar hours. Therefore, a set of agreed ground rules around work patterns is recommended. Embrace flexible working hours to enhance work-life balance, wellbeing and performance, but be cognisant of those with different work patterns and manage them accordingly.

## Adoption of agile working

The adoption of agile working has gradually increased over the last 30 years, with consulting and technology companies being early adopters. Agile working was popularised by Government and banking with academia, R&D and legal practices being late adopters. Its uptake was not wholesale until the Covid-19 pandemic of 2020 that forced the adoption of some form of agile working by those remaining sceptical organisations. Surveys conducted during the pandemic, such as BCO (2020) in the UK and PWC (2021) in the United States, indicate that the majority of staff intends to return to the office for two to three days per week (and conversely continue to work from home for two to three days per week).[5] Consequently, more organisations explored introducing unassigned desking and reducing the number of desks provided.

The number of desks provided in an agile working environment is based on typical occupancy levels, that is, the utilisation of the space. For example, if only half the desks are occupied at any one time during the working week then it might be assumed that sharing could be introduced at a ratio of 2:1 staff per desk. However, utilisation varies throughout the day, by week and season, and between teams so usually contingency is built in. Therefore, a target occupancy of 80% might be set and a sharing ratio implemented of 1.6:1 staff per desk.

Pre-pandemic adopters of agile working typically implemented desk sharing at a conservative 10:8 (1.25:1) or a more ambitious 10:7 (1.4:1) staff per desk. These figures are averages across the organisation with some functions, such as Finance having lower ratios and other departments such as Sales having much higher sharing ratios. Based on a return to the office for two to three days per week, providing unassigned desks for say the 60% of the workforce in the office at any one time is equivalent to a desk-sharing ratio of 10:6 (1.7:1) staff per desk.

Historically, the appropriate desk-sharing ratio was calculated using data from space and time utilisation studies. Basically, prior to implementing agile working, either observers or sensors are used to monitor how often the desks and other work-settings are used over time. The survey may only last one or two weeks if observers are used or longer if sensors are in place and provide a snapshot of how the space is currently used. With the increasing adoption of home working, a mixture of utilisation studies and staff surveys on preferred time in and out of the office is recommended.

Agile working environments usually include multiple work-settings. When desk sharing, for extra contingency, a trick of the trade is to ensure that the seats at the alternative work-settings make up the shortfall between the staff and available desks, sometimes referred to as overflow seating. So, if all staff in the organisation do turn up on the same day, which is very unlikely except maybe for a special event (when they would not be sitting at desks for much of the day), the staff will be able to find a space to touchdown.

Most of the respondents of surveys conducted during the pandemic indicated that they are more productive WFH, this is partly due to being able to concentrate without interruption from colleagues or managers and spending less time travelling to/from the office and between meetings.[6] Nevertheless, feedback also highlighted that creativity and innovation through face-to-face workshops and brainstorming sessions was suffering. Workers wish to return to the office to meet colleagues to facilitate collaboration, brainstorming, social interaction, mentoring and connection.

While the office may be primarily used for collaboration and the home for focussed work, some survey respondents acknowledged that they are struggling to focus at home due not having a dedicated workspace offering sufficient privacy, so they are distracted by family or flatmates. Others said that they did not have adequate furniture and technology at home. The latter is more easily fixed, but the office is required by some to conduct work requiring focus and concentration.

With the increased adoption of home working during Covid-19, some organisations consider the need for less workstations as an opportunity to reduce their office space and save on property costs (assuming their lease terms allow it). In contrast, others are exploring using the released space to address the balance of workstations against space for collaboration, social interaction and focussed work, see Figure 8.3. Many European offices, particularly, those in centres of commerce, are densely occupied due to the high cost-base. Reducing desk numbers, one possible outcome of the pandemic, provides a golden opportunity to reduce the occupational density, space out the workstations and thus make the layout more conducive to

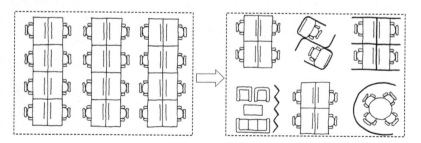

*Figure 8.3* Alternative work-settings instead of desks.

wellbeing and performance by diminishing noise distraction, providing more personal space and helping to reduce airborne cross-infection.

The *Landscaped Office* has a lower density and a higher proportion of alternative work-settings than typical open plan working environments, see Figure 7.7. The ALO, encompassing desk sharing, will therefore be more attractive to those organisations overly concerned with space and property costs.

## Notes

1 Chiat/Day (now TBWA/Chiat/Day) was heralded as a trail blazer for implementing hoteling in its Los Angeles offices. Chiat/Day expected the staff to frequently work outside of the office, but they came in daily and struggled to find a desk. That plus headcount growth and technology issues resulted in reverting to a more traditional office.

2 I am fortunate as I have a home office in which to work but it's actually a wooden cabin, or glorified shed, at the bottom of the garden which allows me to work free from the distractions of the home. I share this workplace preference with Roald Dahl's writing hut, George Bernard Shaw's rotating shed, Le Corbusier's cabanon and all the other shedworkers mentioned by Alex Johnson (2010). Others are less fortunate and make do with working from laptops on their kitchen table, etc.

3 Most definitions of agile working focus on working practices and culture and do not include the concept of hot-desking – that was added by the property industry to help regulate space.

4 Martini is an Italian vermouth. During the 1970s and 1980s, the advertising campaign portrayed a glamorous lifestyle with footage accompanied by the song "anytime, anyplace, anywhere, there's a wonderful taste you can share, the bright one, the right one, that's Martini".

5 The preferred number of days back in the office post-pandemic varied, with a small percentage wishing return to for five days per week and a similar percentage preferring one or even no days.

6 The perceived increase in performance at home may be related to completing transactional work, with short-term objectives and deliverables, whereas long-term productivity may be more relational, comprising creativity, innovation and planning which benefit from the dynamics of face-to-face interaction and non-verbal communication that are often lost in virtual interactions. The latter core work activities are better supported by the office.

# 9 The great indoors

The proposed office solutions presented in the previous chapters mostly relate to the layout and design of the workspace. Indoor environmental conditions, such as acoustics, lighting, temperature and indoor air quality, are equally as important. These hygiene factors are a basic human need that directly affect wellbeing and performance. However, the appropriate conditions will be different for each individual, depending on the activity they are conducting as well as personal factors. Consequently, when groups of people occupy the same space it is particularly difficult to create optimal environmental conditions without offering individual control. Fortunately, research has revealed the ranges of environmental conditions that will satisfy the majority of a group depending on their activity and group factors. Nevertheless, studies continually show that indoor air quality, noise and temperature are found to be the biggest causes of dissatisfaction with workplace design. This is often the result of misunderstanding human needs and preferences, the specific situation, ill-informed design, poor maintenance or simply due to cost-cutting.

Large databases of post-occupancy evaluations, such as Leesman's (2019) database of 5,000+ workplaces and my own modest database of 100 offices, consistently show that the design elements that office workers are most dissatisfied with, and often causing impaired performance, are temperature, acoustics and air quality. Rather than blame the engineers and architects for poor design, let us recognise that the real issue is that these environmental conditions are the psychophysical factors that are highly specific to personal differences.

The following sections are based on Ecophon's *Design Guidance on Reducing Office Noise: A Psychoacoustic Approach* (2021), co-authored by Paige Hodsman and myself, and Sharp's *Creating the Perfect Meeting Environment* (2019) co-authored by Chris Parker and myself. I am grateful to Saint-Gobain Ecophon and Sharp for granting permission for me to rework the original material and present it here.

DOI: 10.1201/9781003129974-12

## Acoustics and noise

### Acoustic standards

As with other environmental parameters, international standards and national guidance exists for acoustic levels in offices. For example, the *CIBSE Guide A* (2015) suggests a maximum background level of NR35 for open plan offices. Sound Pressure Level (SPL) is readily measured in dBA (A-weighted decibels), but as dBA cannot be used to compare sounds across different audible frequencies, Noise Rating (NR) curves are used to convey the frequency information in a single index. The British Council for Offices also uses NR, in their *BCO Guide to Specification* (BCO, 2014), and recommends NR35 for cellular offices and NR40 for open plan. The NR is typically the dBA minus 6, so, for the UK open plan offices, a 41–46 dBA maximum background sound level is preferred. The Association of Interior Specialists also recommends a background level of 46 dBA (AIS, 2011), and British Standard BS 8233:2014 recommends 45–50 dBA. WELL recommends a maximum noise criterion (NC) of 40 for open plan offices, equivalent to 50 dBA.

Although a single figure, the derivation of NR and dBA levels nevertheless seems complex. Acousticians regularly debate which is the best acoustic measure, and international standards use a range of other complex acoustic indices. Take the new *ISO Standard 22955: Acoustic Quality of Open Office Spaces* which recommends measuring five factors: (i) the workstation noise level ($LA_{eq, T}$), (ii) the in-situ acoustic attenuation of speech ($D_{A, S}$) between workstations and across the floorplate, (iii) the A-weighted SPL of speech at a distance of 4 m from the sound source ($L_{p, A, S, 4m}$), (iv) the reverberation time ($T_r$) and (v) the spatial decay rate of speech ($D_{2S}$). Different levels of each acoustic parameter are recommended, depending on the main activity conducted in the open plan space (see Table 9.1). WELL also refers to reverberation time and recommends a $T_r$ of 0.5 s.

*Table 9.1* Acoustic parameters for open plan offices, derived from ISO 22955

| Acoustic metric | Mostly collaboration | Little collaboration | Mostly external comms (phone) |
| --- | --- | --- | --- |
| At the workstation: $LA_{eq,T}$ | <52 dBA | <48 dBA | <55 dBA |
| Between workstations: $D_{A,S}$ | ≤4 dB | ≤6 dB | ≤6 dB |
| On floorplate: $T_r$ | ≤0.5 s | ≤0.5 s | ≤0.5 s |
| On floorplate: $D_{2,S}$ | ≥8 dB | ≥7 dB | ≥7 dB |
| On floorplate: $L_{p,A,S,4m}$ | ≤48 dB | ≤47 dB | ≤47 dB |

Both $D_{A,S}$ and $D_{2S}$ refer to speech/talking, rather than other background sounds, because speech is recognised as one of the main causes of distraction in offices. Speech is also more distracting when it is intelligible, as we unconsciously listen-in and attempt to interpret the speech. Research studies have found that people are less distracted by those around them speaking a foreign language. Many acoustic standards refer to the Speech Transmission Index (STI) and Speech Intelligibility Index (SII) and focus on reducing speech interference. Of course, in an auditorium or meeting space, the opposite is needed, and we require an STI > 0.75 or speech clarity ($C_{50}$) of >6 dB.

### Sound versus noise

The new ISO standard recognises the different requirements for work activities, which is progressed beyond previous standards, but like other standards, it does not readily recognise personal differences and individual preferences. I particularly take issue with guidance and standards intermingling the term "sound" with "noise" in acoustic indices, for example, NR, because from a psychophysical perspective noise it is defined as "unwanted sound".

My extensive literature review of noise (Oseland and Hodsman, 2017) revealed that, typically, just 25% of perceived noise annoyance is attributed to sound level, whereas personal and psychological factors account for up to 75%. This is not surprising as the research literature shows that a particular sound is considered a noise depending upon the context (time of day, source, type of sound, etc.), activity, perceived control and other personal factors such as personality and age. The complex acoustic indices of standards are a good starting point but cannot fully resolve noise issues in the office.

My joint literature review and survey-based research, carried out with Paige Hodsman, found that personality profile affects the response to noise. We used the Big Five Personality Inventory, or OCEAN, to measure: Openness, Conscientiousness, Extroversion, Agreeableness and Neuroticism. We found that people who are more introverted, more neurotic (nervous, apprehensive), less agreeable (less empathetic towards colleagues) and, to some extent, more conscientious (diligent) are more adversely affected by noise in open plan offices. The guidance presented here focusses on introversion/extroversion as this personality trait is generally more recognised and more researched than the other types.[1]

### Solutions for noise

The overarching guidance on office noise, proposed by Paige and myself, can be summarised and easily remembered as DARE: Displace, Avoid, Reduce and Educate.

- *Displace* the noise by providing easy access to separate areas for noisy activities to take place, for example, informal meeting areas, breakout and brainstorming rooms. Ensure that such spaces are located away from the main work areas.
- *Avoid* noise by allowing the occupants the choice of where to work so they can find quiet spaces, for example, focus rooms and WFH. Locate noisy teams together and away from the quieter teams or designate quiet zones within the office where people can work free from distraction.
- *Reduce* the noise distraction by providing a reasonable desk size and limiting the occupational density. Use good acoustic treatments, such as sound-absorbent materials, to reduce speech intelligibility across open plan areas and noise transference between rooms.
- *Educate* by introducing a form of office etiquette, covering phone use, loud conversations, music, headphones, managing interruptions, how different work-settings are used and appropriate use of "do not disturb" signals. Explain to staff how the office layout works, the facilities available to them and how they can control noise.

Our evaluation methodology and DARE principles result in tackling office noise by applying four layers of solutions: Acoustic, Activity, Zoning and Behavioural.

### Acoustic layer

Our approach to resolving noise distraction is a psychophysical one, and this starts with good basic acoustic treatment of the workspace to reduce speech propagation – mainly achieved through absorption, diffusion and sound barriers.

Acoustic ceiling tiles, for instance, will absorb much of the transmitted sounds. A current trend in office design is to have an exposed ceiling slab; whilst this is aesthetically pleasing, it will cause sounds to reflect and travel over distance. In such circumstances, consider installing acoustic ceiling baffles or preferably "rafts"; they come in a range of colours, shapes and designs mostly only limited by our imagination, which can enhance the overall workplace design.

Similarly, wall panels can be added for absorption to reduce speech travelling across the space, particularly, when the ceiling height exceeds 4 m, as well as in conference rooms, to aide clear speech and comfort. Where there is limited space, free-standing acoustic screens or furniture can be used, and panels can be hung vertically from the ceiling. This has the added benefit of breaking up the open plan space or separating equipment areas and breakout or informal meeting spaces, etc. Again, there is an unlimited range of design options for acoustic screens and panels, and they can also be moulded or printed for visual impact and designed to blend in.

Desk screens are also an important means of absorption, especially for teams spending much of their time making telephone or video calls. Ideally, desk screens should be placed between those sitting opposite each other and at a height sufficient to hide the mouth, typically 1,400 ± 100 mm, while still allowing visual connectivity (Figure 9.1). The more fashionable lower screens offer some control over personal space but have little acoustic benefit; side screens (as in cubicles) can lead to reduced communication, loss of perceived space and possibly, isolation. Tallish (1,400–1,600 mm) cabinets, bookshelves or other taller barriers, such as meeting pods or focus booths, placed between blocks of desks will also reduce sound/speech transference and improve acoustic privacy. Also consider the layout of the desks and whether occupants can face away from each other rather than sit face to face.

The *Landscaped Office* is a lower density workplace than most modern open plan offices. A lower density, more spacious, workplace also helps reduce noise. The trend for smaller desks simply means that the occupants are closer together and more likely to be distracted by their neighbours' speech (unless all agree to sit in silence and not use phones or alternatively wear headphones). The trend for increasing space efficiencies, synonymous with higher desk densities and fewer alternative work-settings, results in noisy activities being carried out in the main desk area.

Enclosed spaces, whether a focus/quiet or meeting room, need to be designed for acoustic privacy. Poor sound treatment of such rooms may result in them being sources of distraction. Invest in well-designed acoustic partitions and, if possible, build the partition up to the ceiling slab (not the underside of the ceiling tile). If this is not feasible, consider adding rigid acoustic barriers above the ceiling tiles and wall partitions to stop sound transference over the partition.

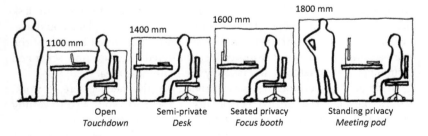

*Figure 9.1* Screen heights and acoustic privacy.

The floor treatment also needs to be considered. Carpet offers more absorption than hard flooring surfaces such as tiles or wood; at minimum, consider rugs in areas where hard flooring is used. Carpet on walkways reduces distraction from footstep noise, and carpet in meeting spaces will reduce noise transference.

## Activity layer

In the design of any office, the work activities carried out by the occupants must be considered so that a range of work-settings can be introduced to best suit those tasks. This approach to designing workplaces is sometimes referred to as agile or activity-based working but is also fundamental to the *Landscaped Office*. For instance, if the organisation or team holds a high proportion of necessary meetings, then provide the appropriate proportion of meeting spaces to support them. However, also consider the type of meeting and relevant type of space, for example, facilitating a productive personal 1:1 meeting requires a differently designed space to a group brainstorming session.

The *Landscaped Office*, and any successful office, has a wide range of readily accessible work-settings, some enclosed and some more open plan, as described in Chapter 7. However, it is not only the visual design and layout of the work-settings that are important but also the acoustic design. If the work-setting is to facilitate noisy activity, then that noise needs to be contained. In contrast, if the work-setting supports focussed or quiet work, then it needs to reduce noise ingress.

## Zoning layer

The location of the work-settings also requires consideration. Far too often in an open-plan office, the work-settings facilitating noisy activities are co-located with work-settings supporting quiet activities. Potentially, noisy spaces, for example, breakout, should not be placed directly next to areas where focussed work is being carried out, quite often the team's main desk area, unless a good acoustic barrier is provided. The workplace could also be zoned according to the potential level of noise generated in each of the work-settings, with noisy work-settings clustered together and separated from quieter work-settings and possibly the main desk area.

In parallel to considering work activity, we also need to create workspaces that support a range of personality and sensory types. Whilst zoning applies to location of the work-settings, it is more relevant to the core activities of the teams and the proportions of personality types within those teams. As mentioned, introverts and the more anxious (neurotic) find noisy

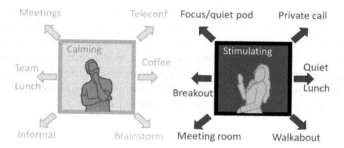

*Figure 9.2* Calming or stimulating base zone.

environments more stressful than their counterparts leading to poor well-being and loss of performance. In contrast, some personality types such as extroverts find noisy environments stimulating, which depending on the task they are carrying out, can improve their productivity. If the balance of personality traits in each team is known, then the appropriate acoustic environment can be created for the predominant type, so either a quiet or noisy base zone (Figure 9.2).

Zoning according to personality type may on first appearance seem impractical, with a preference to co-locate team members. However, several authors have proposed that certain personality types are attracted to particular roles, so it is likely those discipline-based (functional) teams will have a high proportion of the same personality type. For example, a finance team involved in heavy processing of data or a group of analysts will usually attract a high proportion of introverts who are better suited for conducting detailed and repetitive tasks. In contrast, a sales team will attract more extroverts who seek stimulation and take more risks, focus more on the big concept than details and thrive on meeting and socialising with people. It seems counter-productive that many organisations use personality profiling to fit their recruits to their ideal role but then locate them all in the same large open plan workspace even though they require quite different environments to succeed.

Of course, many organisations have multidisciplinary teams based on projects or accounts. Previous research has indeed shown that the most successful and productive teams usually have a rich mix of personality types. In such cases the team members occasionally need to come together to collaborate. Offering choice over where to sit – another core agile and activity-based working principle – allows different personality types to co-locate when required but then they can retreat to the workplace zone that best suits them and their core work activities.

Creating a calming or stimulating base zone is one of the early stages of the *Landscaped Office* design process. In a calming zone, most of the work may involve concentration and require quiet, but at other times the worker will be involved in meetings and presentations or wish to break from their work and socialise with colleagues. So, spaces around and accessible from the quiet zone will be needed for noisy activities. Similarly, those in a stimulating base zone will occasionally require access to quieter spaces for concentration, contemplation or high confidentiality.

*Behavioural layer*

Even the best designed workspace will not be successful if it is inappropriately used. This final layer is less about the physical design and more about how to influence the behaviour of those using the space.

i    *Choice and control* – A first step to changing behaviour is to provide a genuine choice of work-settings that support a range of personality and sensory types carrying out different work activities. These settings should be within the workplace, as part of the ABW menu, but also outside of the workplace such as WFH or remote working in cafés and libraries, etc. Research shows that perceived control is critical in reducing noise distraction. A range of different work-settings offers people perceived control, and this relates as much to how people use them as well as their design. Choosing the time of day to work, whether early or late, and where to work, including occasional WFH, reduces noise distraction. Clearly demarcating the noisy and quiet spaces within the office will inform choice and control. Headphones may also be used to mask noise, but the downside is lost acoustic connection with colleagues.

ii   *Acoustic etiquette* – The occupants may require reminding or educating on acceptable behaviour within a predominantly open plan office. Any guidance or a charter on how to use the office is best generated and agreed by the occupants. Whilst some occupants may find this approach trivial, some personality types (agreeable, introverted, neurotic) welcome some structure and agreed guidelines to refer to. As such, the guidance may include advice on respecting people's time and recognising when they are busy and should not be interrupted. It may include advice on how to tackle people making unnecessary noise through chat, or impromptu meetings, and telephone calls (in particular mobile phones and hands-free). Introducing acoustic etiquette tends to be less successful when a choice of work-settings is not provided.

iii  *Signals and visual cues* – Acoustic etiquette involves reading whether colleagues are busy or available. This usually relates to posture and observed concentration, but some organisations have experimented with using coloured flags or other devices on their desks. An electronic version of "do not disturb" traffic light systems are the presence indicators on instant messaging. Many people use headphones or ear buds to simply indicate that they are busy rather than to actually play music to mask the noise. Our behaviour in an environment is affected by our expectations and experience of that place such that certain design elements can invoke a particular behaviour, that is, Barker's (1968) behavioural settings. To clarify, we all know how to behave appropriately in a library (or church), and introducing the design elements of such environments into the workplace can influence the use of space. One solution is bookshelves and books that not only offer sound barriers and absorption but also invoke quieter behaviour.

Table 9.2 highlights specific acoustic solutions, and work-settings, that are more appropriate for reducing noise for introverts and extroverts. There is some corresponding overlap with sensory avoiders and sensory seekers.

*Table 9.2* Acoustic-related solutions for introverts and extroverts

| Acoustic solution | Introvert | Extrovert |
| --- | --- | --- |
| Planting and moss walls | Calming effect, restorative and helps mitigate noise | Offers interest and aids wellbeing |
| Soundscape – water | Could be calming but also may be distracting | Offers interest, variety and can be stimulating |
| Music | Can be distracting | Can offer interest and stimulation |
| Ceiling absorption | Quality ceiling tiles help reduce noise distraction | Quality rafts help reduce noise distraction to others |
| Wall absorption | Quality wall panels help reduce noise distraction | Quality wall panels reduce excessive noise in meetings |
| Partitions (glazing, to soffit) | Solid panels provide visual and acoustic privacy | Glazed panels in meeting spaces reduce distractions to others |
| Semi-partitioning (open bookshelves, etc.) | Provides some visual and acoustic privacy | Breaks up the space adding some variety and interest |
| Floor covering | Carpets throughout help mitigate noise distraction | Hard floor coverings in social spaces add buzz and interest |

(*Continued*)

| Acoustic solution | Introvert | Extrovert |
|---|---|---|
| Furnishing | Prefer heavy/soft/domestic furniture for calming effect | A mixture of furniture adds variety and interest |
| Large open-plan layout | Prefer single or shared offices or cubicles | Large open plan creates buzz, interest and stimulation |
| Large desk clusters | Prefer singles, pairs or small clusters | Large desk clusters better for interaction |
| Touchdown desks | Can be disorientating and too distracting when used by others | Facilitates interaction with colleagues from other teams and locations |
| Desk screens/ partitions | Prefer higher desk screens (>1,400 mm) for privacy | Lower desk screens better facilitate team interaction |
| Open (informal) meeting spaces | Prefer enclosed meeting spaces | Nearby informal meeting spaces support interaction and tacit knowledge |
| Focus pods/carrels (one-person) | If the main space is open plan with low screens then focus pods appreciated | Some may be required for occasional focussed work |
| Booths/banquette seating | Prefer enclosed meeting spaces but semi- enclosed banquette better than open | Useful as support small impromptu meetings |
| Breakout area | May prefer small tea-point or quiet/reading space | Appreciated as used for impromptu business and social interaction |
| Social/games area/ drinks | Possibly include quieter area for solo or 1:1 activities | Appreciated as used for impromptu social interaction |
| Amphitheatre and brainstorm area | Prefer more formal meeting spaces | An appreciated area for sharing knowledge and ideas |
| Chill/relax area | An appreciated space that supports reading, relaxing and quiet activities | Will be occasionally used for re-energising |
| Library/reading area | An appreciated space that supports reading, relaxing and quiet activities | More likely to read in the breakout and social spaces |
| Enclosed print/ copy area | Would prefer fully enclosed but not too far from desks | Could be semi-open and if away from desks provides opportunity for interaction |
| Connecting staircase | Used but not priority | Ideal for connectivity and impromptu interactions |
| Route through work stations | Would prefer clear circulation away from desks to reduce distraction | A circuitous route offers opportunity for interaction with colleagues |

# Thermal comfort

## Temperature and performance

Seppänen, Fisk and Lei (2006) at Helsinki University of Technology, Finland, conducted a meta-analysis of 24 robust studies of the impact of indoor temperature on performance. Later, Lan, Wargocki and Lian (2011) plotted the relationship between temperature and the performance of typical office workers in temperate climate zones using the Finnish analysis and other research (Figure 9.3). The chart indicates an optimal temperature range of 21–23°C and a decrease in performance of approximately 1% for every 2°C below that range and 2% for every 1°C above it. Other studies have found larger decrements in performance (up to 6%) outside of the base comfort range, and the comfort range also varies between studies and climate zones.

Numerous studies of thermal comfort have been carried out by David Wyon and colleagues over the years. For example, they found that typing comprehension and memory recall are adversely affected when the temperature is 4°C higher or more than that considered optimum for comfort (Wyon, Anderson and Lundqvist, 1979). Memory was also found to be affected by temperatures below that required for comfort. They conclude that "moderate heat stress, only a few degrees centigrade above the optimum, has a marked effect on mental performance when temperatures rise slowly", whereas "memory and creative thinking, are improved by exposure to a few degrees above thermal neutrality, but they too are impaired at higher temperatures". Wargocki and Wyon (2017) conclude that "thermal and air

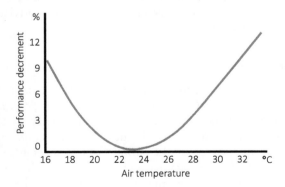

*Figure 9.3* Effect of temperature on performance (derived Lan, Wargocki and Lian, 2011).

quality conditions that the majority of building occupants currently accept can be shown to reduce performance by 5 to 10%". However, the study also illustrates how the optimal temperature varies according to the activity.

Valančius and Jurelionis (2013) found that a short-term temperature drop from 22°C to 18°C increased employee performance by 4.1%. They suggest that "from the point of view of long-term performance, indoor air temperature should be fixed … On the other hand, air temperature should be gradually decreased to 18°C one hour before the end of the working day…". Whilst this is contentious and a unique finding, if verified, it may mean that a short-term drop in temperature would help improve performance in, say, lengthy meetings or long working days.

Indoor temperature not only affects performance but also mood, fatigue, physiology and health. Lan et al. (2011) found that experimental subjects performing neurobehavioral tests and typical office work tasks reported an increase in ill-health symptoms when feeling too warm. Furthermore, the subjects experienced more negative moods, lower vigour and more fatigue when completing tasks under higher temperatures. The performance of the simulated office work and of the neurobehavioral tests also decreased in the warm condition. Physiological measurements were also affected implying that the negative effects due to people feeling too warm are caused by physiological changes. A later study by Lan and colleagues (2020) found that performance was degraded at prolonged moderately elevated temperatures despite the participants maintaining their thermal comfort, due to underlying physiological responses.

Tanabe, Nishihara and Haneda (2007) found that, at higher temperatures, their subjects complained more of mental fatigue, and they found that more cerebral blood flow was required to maintain the same level of task performance. They also found that performance during simulated office work, that is, multiplication and proof-reading tasks, correlated strongly with thermal satisfaction.

### Thermal comfort standards

As with acoustics, there are international standards and national guidance on temperature in offices. Consider the *BCO Guide to Specification* which suggests a minimum temperature of 20°C in winter and a peak temperature of 24°C in summer for offices, but also notes that a 2°C tolerance should be allowed. For all office types, the *CIBSE Guide A* recommends 21–23°C in winter and 22–25°C in summer and acknowledges that temperatures above 25°C adversely affect performance, and offices should not drift above 28°C for more than 1% of the occupied time. International standard *ISO 7730* (2005) provides examples of temperature design criteria for offices. For

Class B offices (providing 90% satisfied occupants), the standard calculates 22 ± 1.5°C in winter and 24.5 ± 2°C for summer. The American standard *ASHRAE 55* (2017) indicates comfort temperatures of 21–25°C in winter and 24–27°C in summer (at 55% humidity). WELL recommends using *ASHRAE 55* for mechanically ventilated offices and the adaptive comfort model for naturally ventilated ones.

It should be noted that the temperatures proposed are operative temperature. Operative temperature is an average of air temperature and mean radiant temperature (measuring heat radiated from surfaces) and is more representative of the temperature that humans experience than air temperature alone. I was once called in to investigate why staff were feeling cold, and wearing scarves and jackets, in their office when the air temperature was measured at 20°C. The office was located underground in the middle of a large archive, and I measured the radiant temperature to be 14°C so the staff were experiencing 17°C operative temperature, which is too low for sedentary activity.

Thermal comfort is not only dependent on temperature but also on other related environmental variables. Air velocity affects thermal comfort, the higher the velocity the lower the perceived temperature, akin to the wind-chill factor outdoors. Office occupants sitting directly under ventilation grills often complain of discomfort as cold air is dumped on them from above. One difference between a breeze and a draught, whether from a window or ventilation grill, is the temperature of the air. Humidity also affects thermal comfort, and generally, a dry heat is more comfortable than humid environments. However, dry environments reportedly cause problems for contact lens wearers and, in extreme cases, result in a dry throat and irritating cough, so 40–60% relative humidity is often recommended. Incidentally, 40–60% humidity also reduces the survival of influenza virus, but the impact on Covid-19 is unclear. Thermal comfort is also affected by a person's activity, such that more sedentary activities have a lower metabolic rate (body heat production) than standing or more vigorous work activities. Thermal comfort is also affected by the clothing worn, which changes the amount of insulation offered.

Most thermal comfort standards have adopted the Predicted Mean Vote (PMV) method of determining thermal comfort, developed by Ole Fanger and engineering colleagues at the Technical University of Denmark in the 1970s. The PMV method uses the four physical parameters (air temperature, radiant temperature, air velocity and relative humidity) and two personal parameters (metabolic rate and clothing insulation) to predict the optimum comfort temperature.

The values of the two personal parameters are estimated using look-up tables. As such, in *ISO 7730*, winter clothing is assumed to be 1.0 clo based on underwear, shirt/blouse with long sleeves, trousers/thick skirt, jacket, socks/ stockings and shoes, whereas summer clothing is estimated to be 0.5 clo representing underwear, shirt/blouse with short sleeves, light trousers/skirt, light socks and shoes. The metabolic rate in offices is estimated in *ISO 7730* to be 1.2 met, based on sedentary activity. However, according to *ASHRAE 55*, the metabolic rate in offices can vary from 1.0 met for seated reading to 1.4 met for filing while standing and 1.7 met when walking around. The PMV methodology is most sensitive to the personal variables, because they have the largest effect on the predicted optimal temperature, and yet, they are dependent on look-up tables rather than measurements. For example, removing a jacket increases the optimum temperature by 2.0°C and even the level of insulation of the office chair offers ±0.15 clo, off-setting the optimum temperature by 0.5°C. Similarly, sitting down and reading, rather than conducting "sedentary activity", increases the optimum temperature by 1.5°C.

The thermal comfort standards are based on steady-state heat transfer physics and therefore tend to treat humans as inanimate objects. But of course, when uncomfortable, humans either adapt their environment or adapt their behaviour to compensate. Allowing occupants to change their attire with a relaxed dress code is one of the most effective ways of regulating thermal comfort. It sounds like an obvious solution but is one often overlooked. In the late 1990s, I presented at the annual conference of the American Society of Heating, Refrigerating and Air-Conditioning Engineers (ASHRAE), the professional association responsible for compiling indoor environment standards, such as *ASHRAE 55*. The conference was in Boston in July, and I clearly recall that the pre-conference literature stated that the temperature in Boston is a balmy 28°C but bring along a jumper or jacket as the conference venue is cooled to 19°C! Maybe those speakers feeling a little stressed, raising their metabolic rate, appreciated the cooler temperatures. Allowing office occupants to change their activity is another easy solution, including standing at a desk rather than sitting. Figure 9.4 (based on Oseland et al., 1998) offers a flow chart that illustrates how changing activity and clothing can be used to offset temperature and alleviate discomfort along with changing the air velocity using local fans or openable windows.

There are other personal factors to consider when determining thermal comfort. While the predicted comfort temperature values given in standards are sensitive to changes in metabolic rate due to activity, it also varies by age, weight, lean (fat-free) body mass, gender, menstrual cycle and psychological state such as anxiety and stress. Traditionally, different temperature requirements for female and male office workers may have been attributed to females conducting more sedentary work whilst wearing skirts and court

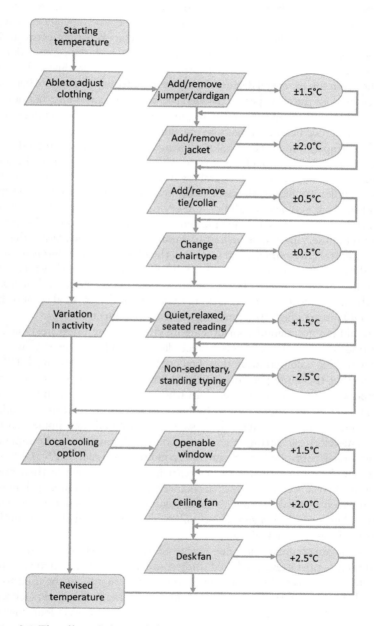

*Figure 9.4* The effect of adaptive behaviour on optimum temperature (adapted from Oseland et al., 1998).

shoes; as the metabolic rate settles down, circulation is reduced to the feet and so exposed legs feel colder. Whilst this is not the case now, differences in average heights and weight will cause differences in metabolic weight between the sexes. Many of us will have experienced an increase in body temperature during a presentation at work, due to increased stress. These individual differences alone can result in several degrees difference in preferred temperature between people and for the same person through the day. Even consuming hot or cold drinks affects metabolic rate. In short, pinpointing the same optimum temperature for a group of office workers is very difficult.

The temperature ranges calculated using thermal comfort standards may also be flawed. The calculations were derived from studies of paid experimental subjects in climate chambers under strict controlled conditions. Such studies, including those of noise and air quality as well as thermal comfort, lack ecological validity. Observing people in artificial environments does not capture how people perceive and behave in the real world. It is akin to observing the behaviour of animals in the old zoos rather than in their natural environment and assuming the observed behaviour is normal. Whilst climate chambers help control the physical variables, they take little account of psychological factors such as motivation, mood and attitude. Furthermore, most experiments do not control for the physical stimuli that are not being studied, thus ignoring multisensory effects.

To aid environmental control, climate chambers are often steel clad; in my research days, I used both a steel clad one and another that was fitted with blue-coloured insulation. Frederick Rohles and Ward Wells (1976) found that the same person in such starkly clad chambers perceived the temperature to be much cooler than when in a furnished chamber of the same temperature; they called it the "meat locker effect". Likewise, I studied the same people in a climate chamber, their office and their home wearing the same clothing, sitting at the same chair and conducting the same activity under controlled environmental conditions (Oseland, 1995). I found that, for the same temperature, my participants felt warmer in their home than in their office and warmer in their office than in the climate chamber. I concluded that this was due to contextual factors such as lighting and furniture, but maybe other sensory stimuli or the emotions evoked by familiar surroundings also took effect.

As climate chambers decrease perceived warmth, then, it is possible that optimum temperatures predicted using standards are higher than required. Contextual factors also imply that non-thermal stimuli can be used to affect thermal comfort. Thus, a more domestic setting in the workplace with subdued yellow spectrum lighting, cosy furniture and richer colours may increase perceived temperature and thermal comfort than a starkly furnished room coloured in pale blue or white.

## Adaptive comfort

Vernacular architecture refers to a type of building design that has evolved over millennia based on local culture, climate, resources and clothing. There are many examples of where vernacular architecture and clothing design symbiotically support the local culture and climate. At one climatic extreme, there are the Inuit people who replace their waterproof sealskin boots, dense polar bear parkas and igloos in winter with soft elk robes, buffalo moccasins and their tupiq tent in summer. In contrast, Arabic clothing includes long flowing robes (the thobe or dishdasha) which create a pumping action with movement to cool the body. Arabic buildings have wind catchers – tall towers that divert cool air to the building whilst pushing out warm air using the stack effect. Thick walls with high thermal mass, shadowed courtyards and landscaping all help keep the occupants cool by natural means.

Japan provides some of my favourite examples of clothing and building design working in harmony. In a ryokan, a traditional Japanese inn, the room is minimalist with a tatami mat flooring, a foldaway futon and a kotatsu, a low wooden table with an underneath heat source. There is no central heating system in traditional Japanese accommodation, so, in bygone years, the occupant would sit in their robe (either a light yukata or a heavier kimono) and pin the corners of the robe to the table so that they received a direct gust of warm air – highly localised rather than central heating.

Modern office design can be quite outlandish compared to traditional vernacular design. For example, the air-conditioned glass towers in the Middle East compared to vernacular buildings with high thermal mass, wind catchers and sheltered courtyards. The modern office basically evolved to accommodate masses of people, paper and technology, but it has also evolved to accommodate contemporary fashion such as the suit. So, maybe, the cultural norm of the 20th century for office workers to wear suits (and ties or scarves) on days when the outside temperature clearly indicates that lighter attire may be more comfortable is partly responsible for the increased perceived need for air-conditioned offices.

Let's consider the suit for a moment. Apparently, the notion of tailoring developed in Europe gradually between the 12th and 14th centuries. By King Louis XIV's reign, the 17th century, men had stopped wearing the doublet, hose and cloak and started to wear coats, vests and breeches, which became the three components of modern male attire. By the start of the 19th century, the upper classes were dressing in a more restrained manner similar to the masses. The suit was born out of tailoring to accentuate the male physique and be less flamboyant than earlier European clothing trends. Of course, females also wear (trouser) suits, but it is less common for male office workers. Regardless, the suit is a fashion item rather than developed out of need, which is quite different to traditional clothing.

The Japanese Government introduced the concept of Cool Biz into the workplace. In Japan's hot summer months, the offices are highly air-conditioned to cool the very formally dressed office workers. The idea of Cool Biz is to encourage office workers to dress down, and for it to become culturally acceptable, so that set-point (thermostat) temperatures in buildings can be raised slightly in summer, thus saving energy and reducing carbon emissions. The campaign has continued to show significant carbon savings year on year. The modern air-conditioned office is undoubtedly seen as a lower risk, lower cost option than vernacular design. Yet, people still prefer an openable window, good daylight and maybe smaller scale buildings. Perhaps, as office workers now tend to dress less formally than previously, we can ditch the suit entirely and encourage developers to build more naturally ventilated offices.

The flow chart (Figure 9.4) illustrates some adaptive behaviours. The thermal comfort standards recognise that people can adapt to higher temperatures, particularly those in warmer climates located in naturally ventilated buildings. As stated in *ISO 7730*:

> In warm or cold environments, there can often be an influence due to adaptation. Apart from clothing, other forms of adaptation, such as body posture and decreased activity, which are difficult to quantify, can result in the acceptance of higher indoor temperatures. People who are used to working and living in warm climates can more easily accept and maintain a higher work performance in hot environments than those living in colder climates. Extended acceptable environments may be applied for occupant-controlled, naturally conditioned, spaces in warm climate regions or during warm periods, where the thermal conditions of the space are regulated primarily by the occupants through the opening and closing of windows.

The adaptive model of thermal comfort uses outdoor temperatures to predict comfortable indoor temperatures for naturally ventilated (free-running) and air-conditioned buildings. Typical monthly outdoor temperatures can be used for design criteria or the running mean outdoor temperature can be used to determine the on-going optimum temperature. The various standards use slightly different equations to calculate the optimum indoor temperature. In the UK, the average maximum summer outdoor temperature is approximately 20°C, and the *CIBSE Guide A* predicts corresponding indoor temperatures of 23.4°C–27.4°C for naturally ventilated offices and

22.4°C–26.4°C for air-conditioned offices. One valid reason for allowing indoor office temperatures to drift is to conserve energy and reduce carbon emissions. More importantly, it allows the occupants to adapt to enhance their comfort and performance. There is most likely an upper temperature limit related to adaptive behaviour (perhaps 28 ± 1°C), so it is unlikely that extreme temperatures tolerated in other climates and cultures transfer to those in temperate climates. Thermoregulation of higher temperatures is likely to be due to ethnic physiological variations, such as differences in eccrine sweat glands (Taylor, 2006), that have evolved over eons.

One inference from the adaptive comfort model is that occupants of naturally ventilated buildings are more tolerant of higher temperatures. This has been backed up with numerous studies comparing air-conditioned offices to naturally ventilated ones. When visiting offices or developing a design brief for my clients, I often hear the occupants asking for openable windows over air-conditioning, maybe because they feel that they have more control over the environment or because they innately prefer natural ventilation. Adrian Leaman and Bill Bordass (2010) often talk of the forgiveness factor in buildings:

> People seemed to be more tolerant of conditions the more control opportunities – switches, blinds and opening windows, for instance – were available to them … People are more forgiving of discomfort if they have some effective means of control over alleviating it. However, many modern buildings seem to have just the opposite effect. They take control away from the human occupants and try to place control in automatic systems which then govern the overall indoor environment conditions, and deny occupants means of intervention.

### Solutions for thermal comfort

Thermal comfort is a psychophysical phenomenon, and therefore, individual requirements depend on several factors. Jerzy Sowa (2020) refers to the Savannah hypothesis, noting that, on the Savannah, such as "Tanzania, the average monthly air temperatures throughout the year are almost constant, between 20.7°C and 25.4°C", and therefore, "regardless of where we live on Earth, people try to maintain an indoor temperature of 20°C to 25°C". Nevertheless, he also notes that "There is no combination of indoor parameters resulting in more than 60% of satisfied people … 20% of people would like the temperature increased, and another 20% prefer reduced temperature". He concedes that differences in body structure, diet, illnesses, etc., lead to different thermal preferences, indicating "the vast potential of decentralized control of HVAC systems".

Despite our innate comfort preferences, part of our survival instinct, and one reason for the success of our species is that we can adapt to our surroundings or we can adapt our surroundings to us. As intelligent animals, we are less likely to sit still and over-heat or freeze, unless of course we are prevented from implementing the adaptive options. In and outside the office, we all have our favourite spots – ones that make us feel comfortable as well as supporting our activities and reflecting our mood. Given the choice, we move to a cooler or warmer spot. In animal enclosures, including my own frog terrarium at home, it is a common practice to provide hot and cool spots for the inhabitants. Unfortunately, in many offices, we do not provide a range of spaces with different environmental parameters but design for the assumed average (based on flawed assumptions and models), and we tend to not allow our workforce to select their workspace based on their comfort.

Back in the early 1990s, I created a spoof article for the *CIBSE Journal* proposing that offices should offer different thermal environments such as the "Hawaiian suite", with a higher set-point temperature and warm/brighter colours, and "Arctic studio", with a lower set-point temperature and cooler colours. I even received some enquiries on how to implement such a workspace but, at the time, it was wishful thinking. The more recent adoption of agile and activity-based working with hot-desking, along with a more sophisticated Building Management System (BMS) and the penchant for "Google"-type funky offices, means that creating different distinct thermal zones in an office is now a more practical and viable solution for the *Landscaped Office*.

In their survey of 11,366 employees located in activity-based working environments, the Leesman survey (Roth, 2017a) showed that a higher percentage of transient workers were satisfied with temperature, indoor air quality and lighting compared to those who sit at the same desk each day. The results indicate that some agile workers select their desk position based on comfort rather than where their colleagues sit. *WELL* recommends that "All open office spaces with occupants performing tasks that require similar workstations allow for at least 50% free address to allow occupants to select a workspace with a desired temperature".

Thermal zones will probably need to be self-contained so need to be wholly or partly partitioned off area on the floors. Creating thermal zones in full open plan offices is more difficult but may be achieved with some separation, perhaps creating a grade of temperatures across the floor. One issue in open plan is that competing requests by the occupants for different thermal environments means that a "soup" of average temperature is created. Furthermore, occupants not receiving their preferred temperature, with no explanation, will be more disgruntled. One way around this is to provide the occupants with a means of voting on the optimum temperature for the group and providing feedback on the chosen temperature.

Many years ago, I helped to develop a system called the Democratic User Control of Zonal Temperature (DUCOZT), which determined the preferred temperature in open plan offices (Oseland et al., 1997). Each time one occupant wanted to change the temperature, a PC pop-up (dialogue box) would be sent to their nearby located colleagues asking if they agreed to increasing or decreasing the temperature. If the majority wanted a change, then a signal was automatically sent to the BMS, and the temperature in that zone of the building was changed accordingly. The occupants then received a feedback pop-up on the current and predicted temperature. Whilst a laudable idea, at that time, there wasn't much uptake on DUCOZT – in hindsight, maybe the system was too costly and difficult to implement. However, it did inspire others to build similar systems. Perhaps it is time to revisit such a system including ones for light and ventilation as well as temperature.

An alternative strategy for the *Landscaped Office* and open plan environments is to use personalised environmental control systems such as the Environmentally Responsive Workstation (ERW), Personalised Environment Module or Climadesk, see Figure 9.5. Yang et al. (2021) provide a review of the various systems available, but they are all basically desks that provide local/personal control of environmental conditions. Most have some form of heating, usually, an under-desk radiant panel, and cooling via desk-mounted fans and vents, and some are also "plumbed" into the air conditioning; the units also have lighting. All systems are connected to an occupancy sensor so that the heating, cooling, lighting and desktop power go into sleep mode when unoccupied. The workstations are more expensive than a standard desk so there has not been wholesale take-up. However, a study of the West Bends Mutual Insurance company's headquarters found

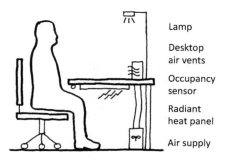

Lamp

Desktop
air vents

Occupancy
sensor

Radiant
heat panel

Air supply

*Figure 9.5* Environmentally responsive workstation.

that ERWs were both economical over time, due to the energy-saving mode, and also increased performance, that is, the insurance claims processed by 2.8% (Kroner, Stark-Martin and Willemain, 1992). I have visited several open plan offices where the staff have installed their own desk fans and under-desk fan heaters. This self-creation of localised control is an example of adaptive behaviour and an extreme form of Bring Your Own Device (BYOD).[2] Facilities managers frown upon such behaviour as there are safety implications, especially, if the equipment has not been Portable Appliance Testing (PAT)[3] tested.

It is worth noting that overly dense working environments can also create issues with temperature control. The cooling and ventilation systems in air-conditioned offices are based on a predetermined density that should not be exceeded. Similarly, naturally ventilated offices may struggle with excess people and equipment.

## Indoor air quality

### Indoor Air Quality regulation

Good air quality is a more important basic requirement than temperature. Alker (2014) of the World Green Building Council (WGBC) states that:

> within a certain temperature range … there are not the same direct risks to health that poor air quality brings. In fact, studies have shown that humans are remarkably adaptable to temperature in a way that they are not, for example, to air quality.

Air quality refers to the level of pollutants in the air, including Volatile Organic Compounds (VOCs), off-gassed by some furniture and building materials, and carbon dioxide ($CO_2$) exhaled by people.

$CO_2$ is often used as a proxy measure of poor air quality because it is a common pollutant, and by maintaining low $CO_2$ levels, then, other pollutants are likely to be reduced. Extreme levels of $CO_2$, fortunately not usually found in offices, can displace oxygen in the air and, in turn, the blood stream and brain, resulting in symptoms such as hyper-ventilation, rapid heart rate, clumsiness, emotional upset and drowsiness. Indoor Air Quality (IAQ) standards recommended optimal $CO_2$ levels but combatting the build-up of indoor air pollutants requires a regular supply of fresh air through a ventilation system or from natural ventilation (windows) in clean air locations. Standards therefore also recommend the appropriate ventilation rates for buildings.

Typical outdoor $CO_2$ levels are 250–350 ppm, and the recommendation for offices ranges in standards from 350 to 1,000 ppm, typically produced using fresh air supply rates of 10 l/s per person or more. The *CIBSE Guide A* refers to *British Standard BS13779* which, for a "medium indoor air quality standard", proposes a ventilation rate of 10–15 l/s per person and indoor $CO_2$ concentration of 400–600 ppm. The *BCO Guide to Specification* states "A minimum of 12 l/s per person should be provided and it is recommended that at least 10% more air is added to account for meeting rooms and areas of high occupation density". European normative standard *EN15251* proposes, for high quality working environments, a minimum of 10 l/s per person for the office occupants plus additional ventilation to account for pollution levels in the building ranging from 5 to 20 l/s per person. For the highest quality, the $CO_2$ should be no higher than 350 ppm above the outdoor levels, so approximately 600–700 ppm. *ASHRAE 62* uses a similar calculation to *EN15251*, but the total ventilation rate for offices equates to 8.5 l/s per person; there are other maximum limits specified for other indoor air pollutants. *WELL* proposes that ventilation rates comply with *ASHRAE 62* and that the mechanical ventilation or openable windows keep $CO_2$ levels below 800 ppm; it also recommends maximum levels for other pollutants.

The recommended ventilation rates refer to fresh air intake, that is, treated outdoor air. However, to save on the energy required to heat outdoor air in winter and cool it in summer, it is common practice in some countries to recirculate a proportion of the air using the mechanical ventilation systems. In the past, such practice was partly responsible for Sick Building Syndrome (SBS), and if the recirculated air is not sufficiently treated/filtered, it could spread airborne diseases and other pollutants. Monitoring the parts per million of $CO_2$ and using it to inform the BMS is a more reliable way of ensuring fresh air is supplied rather than depending on the ventilation rate *per se*. A related strategy to help reduce airborne cross-infection is to purge or flush the building by increasing the ventilation rate for a short period overnight.

## IAQ and performance

The WGBC (Alker, 2014) reviewed 15 studies linking improved ventilation with up to 11% gains in productivity, due to increased delivery of fresh air to the desk and reduced levels of pollutants. One office simulation found that increasing ventilation rates from 5 l/s to 20 l/s per person and reducing VOCs improved performance by up to 8%. Another lab-based study found that $CO_2$ had a detrimental impact of 11–23% on decision-making tasks at 1,000 ppm compared to 600 ppm despite 1,000 ppm being considered acceptable in offices. Finally, a comprehensive analysis by Carnegie

Mellon concluded that natural ventilation or mixed-mode conditioning could achieve 3–18% productivity gains. Considering all the studies in their review, the WGBC concludes that "a comprehensive body of research can be drawn on to suggest that productivity improvements of 8–11% are not uncommon as a result of better air quality" (Alker, 2014).

Many studies show a negative impact on performance under high levels of $CO_2$. Maula et al. (2017) found that it had a more negative effect on information retrieval, subjective workload, perceived fatigue and lack of motivation, Satish et al. (2012) found a notable decrease in decision-making performance, and Katjár and Herczeg (2015) observed a significant decrement in reading performance. Allen et al. (2016) compared experimental subjects in a controlled office with air quality conditions representing "conventional" (high concentrations of VOCs) and "green" (low VOCs) office buildings in the United States. On average, cognitive scores were doubled in the green offices compared to the conventional one and were affected by both VOCs and $CO_2$.

Others have focussed on the impact of ventilation rate on performance. Wargocki, Wyon and colleagues have conducted and reviewed many studies of air quality. They conducted a series of studies exploring performance tasks for subjects exposed to a pollution source (a hidden old office carpet) at a 10 l/s per person fresh air supply rate. The experimental participants typed 6.5% more slowly, made 18% more typing errors and experienced more headaches under the polluted condition. In an earlier study, Wargocki et al. found that the experimental subjects improved their creative thinking responses by 10% at higher ventilation rates. In another experiment, the memory recall, typing and arithmetic of participants improved by up to 5% when ventilation rates were increased. Furthermore, Wargocki et al. found that field intervention studies confirm their laboratory findings. In particular, the performance of call-centre operators in Denmark improved by 6%, in California by 2% and in Singapore by 9% when the fresh air supply rate was increased at different levels.

A Finnish meta-analysis by Seppänen and Fisk (2006) illustrates the relationship between the outdoor air supply rate per person and the performance of office workers (Figure 9.6). The reviewed studies showed a clear improvement in performance in tasks requiring cognitive activity when ventilation rates increased. Their consolidated results indicate that an increase of 3 l/s per person results in an approximate 1% improvement in performance, but the effect starts to plateau at around 30 l/s per person.

These extensive IAQ studies clearly indicate that increasing the fresh air supply rate in offices, and undoubtedly in meeting rooms which often become stuffy in extended and crowded meetings, will reduce $CO_2$, VOCs and other pollutant emitted by humans, thus improving the performance. Such a strategy requires a well-designed and maintained ventilation system or access to openable windows in suitable locations.

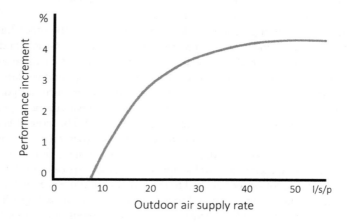

*Figure* 9.6  Effect of ventilation rate on performance (derived from Seppänen and Fisk, 2006).

Research into air quality and ventilation rates tends to focus on group effects rather than individual differences. High $CO_2$ levels cause lethargy through a physiological response rather than a psychological one. Nevertheless, it is possible that those seeking stimulation, like extroverts, may be more tolerant of stronger odours, but there is no clear evidence for it. However, sensory profiling highlights how some people are more sensitive to odours than others, and this is likely to affect their workplace experience, comfort and performance. Furthermore, certain odours can aggravate asthma so would affect some individuals, and hay fever sufferers would not welcome "pollutants" of higher fragrance such as flowers and perfumes.

### Scents and odours

Scents are infused into retail outlets to enhance the shopper's experience, and there is some take-up in offices (Mathews, 2006). According to the manufacturer Signature Aromas, Japanese businesses routinely use their air-conditioning systems to disperse fragrances such as citrus early in the morning to boost concentration, and woody scents like cedar to relieve tiredness in the afternoon. The Takasago Corporation found that 54% of their typists made fewer errors when they could smell lemon, 33% fewer with jasmine and 20% fewer with lavender (Card, 2014). The University of Cincinnati found that volunteers who inhaled peppermint and lily-of-the-valley odours while at work on a tedious computer task made 25% fewer errors than those who breathed unscented air. A recent study in a

UK financial services company concluded that staff made 40% fewer errors when surrounded by the smell of cinnamon.

Despite such findings, the introduction of scents into the office has not had great uptake worldwide, possibly because, for some, like asthma and hay-fever sufferers or those with olfactory hypersensitivity, the "office canaries", introducing odours can have a negative effect on their health and performance. Furthermore, the introduction of fragrances in cleaning products, air purifiers, etc., particularly those concocted with terpene, can react with ozone to produce new undesirable pollutants (Weschler and Shields, 1997). The more direct link between the sense of smell and the brain, compared to other senses, means that odours are highly individualised. It is therefore more common for fragrances to be introduced into confined areas like toilets or breakout space than the whole office. Maybe a scent better matched to the function of the space, such as baking bread or coffee in breakout and social spaces, is more appealing. Of more importance is the removal of offensive odours. An odorous bin in an office or breakout space will certainly impair the overall experience of that space as will people eating pungent food at their desks.

### Solutions for air quality

Everyone wants good air quality, and there is a little variation in individual preferences. The research indicates that higher ventilation rates are good up to a certain limit, and the optimum is higher than specified in some of the more basic standards. In the *Landscaped Office*, the air provided should be "fresh", and ideally treated, with the $CO_2$ levels as close as possible to outdoor levels in unpolluted areas. Increased ventilation will be required in more odorous areas such as toilets, waste, kitchens and breakout areas.

As with thermal comfort, providing more fresh air through windows, or ventilation grills, can result in draughts and some occupants are more sensitive to draughts than others. Therefore, consider who sits near a window and avoid placing desks under ceiling ventilation grills.

The jury is out on introducing fragrances into the workplace but, if used, consider them in selected areas. People like openable windows, especially if providing a gentle breeze, but provision must be balanced against local outdoor pollution levels – consider both air quality and noise.

## Lighting and daylight

### Lighting and performance

There are many lighting metrics, but surface, or desk, illuminance is relatively easy to measure and popular with researchers. In their review of

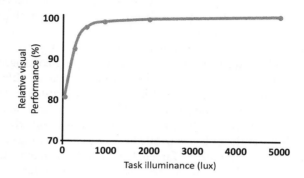

*Figure 9.7* Effect of illuminance on relative performance (derived from Bowers, Meek and Stewart, 2001).

lighting, Attema et al. (2018) calculated a mean increase in performance of 15% due to good lighting, whereas Hiroshi et al. (2006) reported a 9% improvement. Likewise, Bowers, Meek and Stewart (2001) plotted the relationship between the relative visual performance and desk illuminance to find that the performance plateaued at 1,000 lux (Figure 9.7).

Barnaby (1980) studied workers at a life insurance company conducting difficult paper-based tasks. He found that increasing the illuminance from 550 to 1100 lux reduced errors by 2.8% and increasing it to 1,600 lux improved performance by 8.1%. The subjects also rated the higher illumination as less stressful and more motivating. However, Barnaby found that, in the areas where reading was not a priority, the spaces were considered over-illuminated. Based on the research, it appears that 500–1,000 lux task illumination is appropriate in most cases, lowering to 300–500 lux with screen usage.

The studies of de Vries et al. (2018) and others have shown that altering the lighting in a space can change social behaviour in both positive and negative ways. They found that participants in darker environments are more prone to aggression but, in contrast, cooperation and creativity can also be better in dimmer conditions. Lighting is therefore an important consideration when designing for different office activities.

### Lighting standards

Clearly lighting is required to enable work activities to be carried out. The more paper-based activities, such as reading or drawing (on boards), require more lighting than those which are more screen based, such as software developers. One challenge is designing offices that need to facilitate screen activities versus more paper-based activities. Well-designed office lighting is a balanced mix of good desk/task illuminance, ceiling/wall illuminance, ambient lighting and daylight.

The *CIBSE Guide A* recommends a desk/surface illuminance of 300–500 lux for open plan offices. In an earlier version, the guide suggested 300 lux for moderately easy visual tasks, 500 lux for moderately difficult tasks, such as in a general office, and 750 lux for difficult visual tasks such as an old-fashioned drawing office. The *BCO Guide to Specification* states a "target average maintained illumination: 300–500 lux with the ability to enhance lighting at the task area to 500 lux using task lighting or controls". Similarly, *WELL* proposes that "Targeted task lighting can provide the necessary amount of light at workspaces without over-illuminating ancillary spaces; ambient light levels of 300 lux are sufficient for most tasks", and if ambient light is below 300 lux, then task lights providing 300–500 lux should be available upon request. Task lighting has the added benefit of offering local personalised control and compensates for the variation in personal preferences.

The *BCO Guide to Specification* emphasises that "The amount of daylight available to occupants will be seen as an important indicator of workspace quality. Spaces that predominantly rely on artificial lighting will not provide anything like the same standard of lit environment". CIBSE uses daylight factor as a metric of the quality of daylight in an office. The daylight factor is a predictor of occupant satisfaction with daylight and may be used as an initial design parameter. CIBSE state "If the average daylight factor exceeds 5% on the horizontal plane, an interior will look cheerfully daylit, even in the absence of sunlight". In contrast "If the average daylight factor is less than 2% the interior will not be perceived as well daylit and electric lighting may need to be in constant use".

A good daylight factor target for offices is therefore 2–5%, and 80% of the office floor space should have a daylight factor greater than 2%. The quality of the light and the corresponding hue, that is, colour spectrum, are also relevant.

Predicting the daylight factor in an office space is complex and is usually determined through computer modelling and simulation. The location, orientation and elevation of the building and its relation to the sun's seasonal path and other nearby buildings, creating shade and shadows, is key. The size of the window (or other glazed aperture size), glass transmittance and any tinting, window blinds or shading (Brise soleil) are also key parameters along with other factors such as inside surface reflections.

## Lighting and preference

Marija Despenic and colleagues at Philips Lighting provide a comprehensive summary of individual preferences for lighting (Despenic et al., 2017). Guy Newsham and colleagues at the National Research Council Canada have also examined lighting preferences. They found that the preferred desk illuminance was between 83 and 725 lux in open plan (actually cubicles), and the selected illuminance was 116–1,478 lux in a simulated office – quite a range (Newsham et al., 2004). In another field study of office workers who were given control over lighting, Boyce et al. (2006) found the range of preferences to be from 252 to 1,176 lux. Similarly, in a longitudinal study of office workers in four buildings, Moore, Carter and Slater (2003) found that the mean selected illuminance ranged from 91 to 770 lux.

The average lower and upper levels of the preceding four studies are from 135 to 1,038 lux, indicating a range of 900 lux. Figure 9.8 compares the 900 lux range (assuming a normal distribution) with the 200 lux range recommended in standards. Boyce and colleagues calculated that, for a typical office, the maximum number of occupants that would be within 100 lux of their preference is just 65%. Despenic et al. (2017) conclude that "Due to the broad range of individual lighting preferences, it is a challenge to create satisfactory lighting conditions in a multi-user space by providing fixed lighting conditions to all users".

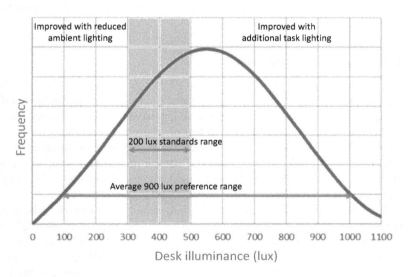

*Figure 9.8* Preferred lighting range versus standards.

Despenic and colleagues also found that:

> when users can adjust the illuminance level on their desks, it has a positive effect on their satisfaction with the environmental conditions, with lighting quantity and quality, mood, improved motivation and vigilance, and indirect positive effects on their productivity ... On the contrary, having a workspace without some degree of control over the environment, leads to increased discomfort and stress.

They go on to develop a method to predict lighting levels based on profiling the occupants' activeness, tolerance, dominance and lighting preferences. They recommend that the profiling is used to create different lighting zones that match the profiles.

Human physiology also affects lighting preferences. It is no secret that our sight deteriorates with age, in turn, affecting the required lighting levels. With age, the lens in the eyes become rigid making it more difficult to focus, ciliary muscles lose strength also making it harder to focus, corneas become thicker and opaquer and pupils shrink so that less light reaches the retina (Park and Farr, 2007). The upshot is that, in general, older occupants require higher light levels to see smaller print and are more sensitive to glare. With increasing retirement ages, there are more older people in the office, often long-standing and senior in the organisation, and we need to cater for them.

### Daylight and colour spectrum

Access to daylight is also important as it directly affects human physiology, health, performance and mood. Natural daylight lies at the blue end of the light colour spectrum, whereas yellow tones, closer to the opposite end of the spectrum, correspond to loss of daylight at dusk and dawn. In general, "warm lights make the environment feel more welcoming and relaxing, while cooler lights make the environment more stimulating – they make us feel more alert, more focused, and can increase productivity levels" (Souza, 2019).

Borisuit et al. (2015) studied people's behaviour under electric light and daylight conditions over several weeks. They found that, in blue-enriched light during the daytime, office workers reported higher subjective alertness, enhanced performance and less sleepiness compared to polychromatic white light. In fact, they discovered that just 30-minutes exposure to bright daylight near windows (>1,000 lux) was as effective as a short nap in reducing post-lunchtime drowsiness. Lee, Moon and Kim (2014) examined people undertaking computer and paper-based reading tasks at low (500 lux) and higher (750 lux) illuminance levels under a range of light colour spectrums. The participants preferred higher colour temperatures at the lower illuminance levels. So, perhaps, slightly bluer light would compensate for lower light levels and help maintain focus and alertness at the office and in longer meetings.

Hormones are produced by glands and then secreted directly into the blood which transports them to other organs to take effect. There are many types of hormones that affect bodily functions and processes, but ones of particular interest to workplace designers are cortisol, serotonin and melatonin. These hormones all help regulate our circadian rhythm, or waking and sleep cycles, and are dependent on daylight (see Figure 9.9).

Daylight stimulates cortisol, also known as the "stress hormone", to be released by the adrenal cortex and so prepare us for action. Serotonin, a neurotransmitter sometimes referred to as the "happy chemical" as it enhances mood, is also stimulated by light. Daylight intensities found in the morning typically produce optimal levels of serotonin to engender a state of high alertness but not stress. In contrast, light supresses melatonin so that when daylight levels fade, the pineal gland produces more melatonin, which in turn induces relaxation and aids sleep.

Light can reset the biological clock and so advance or delay the circadian rhythm. Lack of daylight can therefore affect evening sleep patterns, in turn, affecting alertness during morning work, but absence of daylight may also trigger early (afternoon) drowsiness. The WGBC reported that workers in offices with windows had on average 46 minutes more sleep each night compared to workers without them and being close to windows increased focused work by 15%.

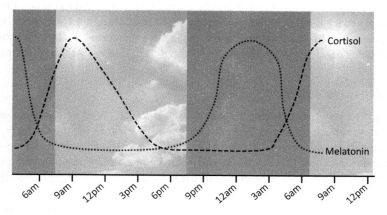

*Figure 9.9* Circadian rhythm.

Cortisol follows a diurnal pattern, with high values during the day and low values at night but also exhibits a seasonal variation with higher levels in summer than in winter. One study found that office workers close to a window had higher levels of cortisol during summer than in winter, making them feel more alert and active (Erikson and Kuller, 1983). So, dark and dingy, or deep plan, office spaces devoid of good daylight could reduce cortisol and serotonin but increase melatonin production, thus inducing drowsiness in office workers. Cool white LED lighting has more blue light than standard incandescent or fluorescent lighting so is better for maintaining alertness during the day. On the other hand, many electronic screens also have blue light, which may prevent sleep if used late at night.

Daylight is also essential for Vitamin D production; it is synthesised in our skin when exposed to direct sunlight. Vitamin D is actually a hormone, and it allows the body to absorb calcium which is essential for strong and healthy bones. Vitamin D deficiency is also associated with muscle weakness, immune system disorders, fatigue and depression. Recent studies have also found a relationship between Vitamin D and cognitive impairment in the elderly, a precursor for dementia and Alzheimer's disease. There does not appear to be a significant effect on daily work performance other than due to fatigue from long-term deficiency. Fortunately, exposure to sunshine for 15–20 minutes for three days per week is usually sufficient for Vitamin D production. Extended lack of daylight in winter, in some regions of the world, or continuous night-shift working is likely to have more impact, but excess Vitamin D can also be stored by the body and used as required.

Whilst a substantial body of research has shown lighting to affect mood, wellbeing and performance, studies of personal requirements are lacking. Photosensitivity, or photophobia, affects a very small proportion of the working population, but such people are occasionally found in offices. Photosensitivity is when a person perceives light much brighter than others resulting in pain and triggering symptoms such as migraine. Bright, blue and flickering light has more ill-effects on them than dimmer warmer lighting.

### Solutions for lighting

Lighting requirements depend on task, with screen-based activities requiring less surface illuminance than paper-based ones. Individual lighting preferences also vary by a considerable range beyond the recommendation in standards. Lower lighting levels with additional task lighting will facilitate office areas with mixed activities in the *Landscaped Office*. It will also offer personal control and help those whose sight is deteriorating with age, or from other causes, plus, it will meet a wider range of individual lighting preferences.

There are clear benefits to good daylight, but the downside is that glare can interfere with screen-based activity, and some office occupants may be more sensitive to glare than others. Those sitting near windows are more likely to shut the blinds if they are sensitive to light and glare. Often, once blinds are down, they remain down, so it is better that those struggling with glare sit away from the windows. Glare can also be avoided by arranging desks perpendicular to the windows or by providing anti-glare screens for computer monitors. Daylight should also be provided in the breakout spaces, perhaps through atria, rooflights and indoor-outdoor type spaces.

Lighting in the meeting rooms of the *Landscaped Office* will need to be adjustable, depending on whether they are being used for a presentation or a brainstorming session. However, avoid dark and dingy meeting rooms by using high luminance displays or projectors for presentations. Generally, non-desk areas, such as breakout space, will benefit from more daylight, but some users of those spaces may prefer more subtle lighting. As with other environmental conditions, consider providing zones in the *Landscaped Office* with varying light conditions and allow the occupants to sit where it best suits them. It is analogous to a natural woodland glade which offers a range of light levels in which people choose where to sit. Electric lighting systems that mimic daylight and vary the conditions (levels and hue) throughout the day are also worthy of exploration.

## Design for all not just the average

After delivering a morning's workshop on how to resolve noise issues in the office by taking account of personality and behaviour, a young acoustic engineer sat at the front asked, "so what number do I use?" Three responses sprung to mind: (i) have you not listened to or understood what I have been saying for the last three hours, (ii) if predicting the required environmental conditions is a simple case of looking up a single number then we do not need professional acousticians, engineers or designers and (iii) Heisenberg's uncertainty principle. In quantum mechanics, Werner Heisenberg (1927) postulated that the more precisely the position of a particle is determined, the less precisely its momentum can be predicted and vice versa. In other words, the more we focus in on accurately measuring a single variable them, the less likely we are to capture other key variables taking effect. A single number cannot represent the complexity of human requirements, no matter how accurate a measure it is.

The impact of indoor environmental conditions on comfort, performance and wellbeing is not just dependent on the task in hand but, as these conditions are all psychophysical ones, our requirements vary according to physical parameters, physiological ones (such as age, gender and sensitivity) and psychological factors (such as personality, mood and attitudes). Furthermore, because they are psychophysical, the requirements will vary over time – not just the season but also the day. To some degree, this makes prescribing single environmental criteria for a whole office a nonsense.

In addition, the environmental conditions recommended in standards are based on average requirements derived from assumed or estimated occupant norms. Many of the standards recommend different precision and ranges of the environmental criteria for several categories of increasing quality. For example, in *ISO 7730*, the temperature requirements are based on three quality levels relating to the percentage of people dissatisfied (PPD), derived from early laboratory experiments and modelling. So, Category A relates to <6% PPD, B to <10% PPD and C to <15% PPD. This, in turn, relates to three temperature recommendations for offices in winter: A = 22 ± 1°C, B = 22 ± 2°C and C = 22 ± 3°C. The idea being the tighter the range, the better the environmental quality. This is fine if the occupants do indeed all have the same physiology and psychology, but if they do not (which is more likely), then, the higher quality categories with the tighter criteria will actually exclude the requirements of even more occupants. The standards are based on engineering precision rather than human imprecision, individuality and diversity.

If the make-up of occupants is fully understood, then the standards can be used as a starting point to determine the ball-park requirements. But these requirements will vary between people and over time. The *Landscaped Office* acknowledges that office occupants have a wide range of comfort requirements and we must design for that range not the average. For any office with its diverse inhabitants, the most viable solution is offering the choice of a range of environmental parameters throughout the office building and/or providing individual control. In bygone years, this would have been impractical, but it is more feasible with the growing acceptance of agile and activity-based working.

## Notes

1   Ambiverts are recognised by some psychologists as those with a balance of extrovert and introvert traits in their personality. Whereas others consider people to lie on an introversion to extroversion scale with a bias towards one end of the scale.
2   BYOD usually refers to staff bringing in their own technology such as handheld devices and laptops.
3   PAT is the process of routinely checking the safety of electrical appliances.

# Epilogue

# 10 Concluding remarks on the *Landscaped Office*

This book offers high-level guidance on how to design and manage a workplace to meet the needs of its diverse occupants. It is a book of two halves. The first part describes the current situation, reflecting on and how the original open plan concept has been misinterpreted to create something akin to a *Workplace Zoo*. It also reviews the psychological literature, identifying the basic human needs that need to be accommodated to enhance worker wellbeing and performance. The second part offers my revived and revised solution – the *Landscaped Office*. As open plan has become synonymous with loathed poor-quality workspaces, the term *Landscaped Office* is an alternative phrase that represents a more promising and considered workplace solution.

## Part 1: Situation – *Workplace Zoo*

The primary purpose of an office is to facilitate the business and activities of the occupying organisation. For an organisation to be successful, for it to perform well, it requires the individuals within its workforce to be performing to their maximum potential. The workplace design, layout, facilities and operation must therefore support the wellbeing and performance of the staff. However, the Corporate Real Estate (CRE) industry often views the workplace as a cost burden rather than an enabler or investment in people. This is confounded by a difficulty in measuring performance compared to the measuring cost. The focus is therefore often on managing costs which constrains the workplace design and limits providing the best facilities for the workforce.

One of the largest costs to an organisation is the property, such that, over the years, the occupational density in offices has increased in order to save on space and property costs. However, the largest cost to an office-based organisation is its people, and coincidentally, the people are an organisation's biggest asset. Small, misguided, savings in property costs could have

DOI: 10.1201/9781003129974-14

a disproportionately higher impact on wellbeing and performance. Cost-benefit analysis is therefore required which compares costs and savings with the less but nevertheless tangible benefits such as increased performance like sales, deliverables and innovation, along with reduced absenteeism, disruption and staff attrition. A small increase in staff performance will likely offset the annual workplace costs.

A misdirected approach to office design and planning over the years has, in many cases, resulted in the *Workplace Zoo*, the modern interpretation of the open plan office. Now, more than ever in a post-pandemic era, is the time to remodel the office to address our psychological and physiological needs. Office design needs to be evidence-based and human-centric to ensure that it meets those basic needs rather than simply ignore them or have a detrimental effect. The psychological and related literature provides ample evidence on how to design offices to meet our innate and evolving human needs. The design should allow office inhabitants to flourish and thrive rather than simply survive.

This book covers human needs and the impact on workplace design based on the following fields of expertise.

- *Psychophysics* – Early psychophysics experiments plotted the relationship between perception and other responses to physical parameters such as light, sound and temperature. Fundamentally, the studies highlighted how our responses to different stimuli are not linear, and they also found that responses vary by individuals but they can be grouped to some extent. The standards for environmental parameters are mostly based on psychophysical studies. The standards either provide recommended temperatures, sound levels, lighting levels and ventilation rates or the means of estimating them for a group of workplace occupants. This is a good starting point, but the psychophysical studies are mostly laboratory-based rather than real world. As such, the standards miss the subtleties and range of responses, behaviours, preferences and requirements for different people under different circumstances in different places.
- *Environmental psychology* – My own discipline, environmental psychology, built upon psychophysical principles but with more emphasis on behavioural responses to complex and multiple variables in real-world studies of the built environment. Environmental psychology highlights how requirements and behaviours vary based on our experience and expectations of a space, situational factors such as the activity taking place and who with, and background factors such as demographics. This not only affects the preferred environmental conditions but also the design and layout of the workplace. For example, much environmental

psychology research has explored the complexity of space and privacy requirements, all relevant to workplace design.

- *Evolutionary psychology* – This is a relatively new field of psychology concerned with our innate psychological needs that have evolved over aeons. Humans evolved to live and survive on the African Savannah, and as such, our environmental preferences relate to natural conditions. This affects our requirements for indoor environmental conditions and the design and layout of the workspace. Biophilia, or an innate affinity to nature, indicates preferences for natural materials and patterns, varying environmental conditions, daylight, natural sounds, social spaces, nooks and crannies, refuge/shelter, etc. Furthermore, greenery through plants and landscaping can be reenergising and aid creativity in the workplace.
- *Anthropology* – Dunbar's Number relates to ideal team and organisational sizes. In turn, this impacts on the layout of office floorplates, creating environments of a more human scale with identified team zones.
- *Personality theory* – Personality is an individual's unique set of traits and relatively consistent pattern of thinking and behaviour that persists over time and across situations. Research shows how personality affects tolerance to sound sources and interpretation of noise along with preferred levels of interaction, meeting spaces, desk type and work location. To maximise wellbeing and performance, the range of personality types represented in an organisation must be considered in workplace design and operation. Certain personality traits are attracted to specific roles such that activity and personality can be considered together. Assuming one-size-fits-all and creating a homogeneous workplace for an assumed average person is not the solution.
- *Motivation theory* – The classic motivation theories of Maslow and Herzberg are referred to throughout the book. For humans to reach their maximum potential, and inferred performance, various needs must be met. This starts with the basic needs, or hygiene factors, such as good environmental conditions, safety and nutrition. It then goes on to higher-order needs, or motivational factors, such as a sense of belonging and accomplishment which are also influenced by workplace design and operation.
- *Sensory design* – Advocates of sensory processing and multisensory design remind us to design for all seven senses rather than simply how the workspace looks. Noise in the workplace is a well-recognised issue that must be considered to avoid loss of performance and wellbeing. However, materials and odours, affecting touch and smell, may have subtle unconscious effects.
- *Inclusivity* – As well as different personalities, the workplace includes valuable staff from different cultures, age, gender, disabilities and

various degrees of neurodiversity. It is better to follow the principles of universal design and design for all as the default. To maximise the wellbeing and performance of all the workforce, we must design for the range and not the average.

The above basic human needs are required regardless of the location that work takes place. However, the emphasis of this book is office design.

## Part 2: Solution – *Landscaped Office*

In the latter part of the book, I provide the foundation for an improved working environment. You may have expected some great revelation where I present my all-encompassing reinvention of the workplace, but that is not to be. The proposed *Landscaped Office* solution builds on the original open plan concept *Bürolandschaft* reinterpreted with a contemporary twist and including elements of the *Action Office*, *Free-range Office* and other office concepts. I have also incorporated long-standing research that is often forgotten or simply ignored. It is an alternative to the repeatedly found homogenous, uninspiring, unhealthy and unproductive offices encapsulated in the *Workplace Zoo*. The revised *Landscaped Office* is also complemented by agile working concepts, creating the *Agile Landscaped Office* (ALO).

There are four fundamental physical components to the *Landscaped Office*, and any well-designed workplace, as illustrated in Figure 10.1 and summarised below.

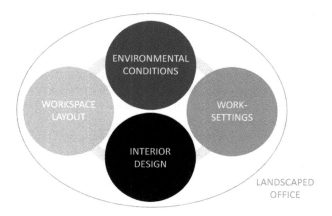

*Figure 10.1* Physical components of the Landscaped Office.

- *Workspace layout* – Open plan offices have received more than their fair share of criticism, mostly from the trade and popular press. The reality is that, like all design, there is both good and bad open plan. Original open plan offices design concepts, such as *Bürolandschaft* and the *Action Office*, had good intentions based around providing an egalitarian and organic layout promoting interaction but with sufficient space and partitions for privacy. Pressures on workspace, prime real estate and associated property costs resulted in a trend for a cheap copy of those original concepts, with overly dense workspaces offering fewer partitions and less facilities. High density and lack of screens is not good for noise, a common problem in open plan, or for airborne cross-infection. Nevertheless, a legacy of poorly designed and implemented open plan working environments does not mean that the original concept does not work or that all open plan offices are sub-standard.

  Having said that, those original open plan concepts can be modernised and improved upon. The desks and alternative work-settings can be planned to create a more organic landscaped feel to the workspace. Different shapes and arrangement of the standard desks can create more interesting, non-orthogonal layouts. The *Bürolandschaft* and *Action Office* had an organic feel, but on appearance, the layout lacked any cohesive structure, and even nature has structure. The uptake of working from home, during the 2020 Covid-19 pandemic, provides the opportunity to review the modern office workspace, address the issue of high density and ensure that the balance of desks to alternative work-settings is appropriate going forward.

- *Work-settings* – Fewer people in the office at the same time means that agile working, with WFH and desk sharing (unassigned desking) when in the office, can be more readily implemented allowing the number of desks to be reduced, thus creating a more spacious office and more space for fixtures and furniture that facilitates collaboration, creativity, concentration and confidentiality. A more generous workspace will provide areas for social interaction, including good refreshments and perhaps games, and zones for quiet work, chilling and contemplation. Implementing shared unassigned desks means that the workplace can be improved without taking on more space and incurring higher costs. But the temptation to implement a combined hot-desking and high-density workplace, driven by cost savings alone, must be avoided.

  As humans have different needs and preferences, it is highly impossible to create a homogenous workplace that suits all. A choice of work-settings and environmental conditions is the solution. Office occupants should therefore be trusted and empowered to work in different settings in and outside of the office. The work-settings offered in the office may be open, semi-open or enclosed; the spaces may have no, low or high partitions such as booths, pods and rooms. Such work-settings can be used to break up the open space creating zones for teams or different activities.

- *Interior design* – The materials, patterns and colours found in nature provide an appropriate look and feel to the space, that is, we should adopt biophilic design principles. Colour, artificial lighting and natural patterns can be used to create different moods such as calming versus stimulating spaces. Visual clues signify the anticipated behaviours in the space, for example, books and bookshelves indicate quiet, whereas coffee points, round meeting tables and games (pool, table tennis) suggest social interaction. As per our biophilic preferences, well-managed planting should also be provided throughout the floorplate, and it also assists with acoustic and visual privacy. The workplace should be varied with areas attractive to all occupants.
- *Environmental conditions* – Our requirements and preferences for environmental conditions also vary, depending on personality, physiology, age, activity, etc. Preferences also vary throughout the day, and innately, we prefer some fluctuation rather than the narrow steady-state conditions recommended in some standards. So again, choice is key. Provide some control over the local conditions, for example, using task lighting, environmentally responsive workstations and quiet pods. Alternatively, provide zones in the office with a range of basic conditions like different levels of temperature and noise. Key environmental factors to account for are noise and thermal comfort, especially in more open plan workspaces or those with exposed ceilings.

Our psychology, physiology and motivation mean that different people are more productive at different times of the day. Choice of when, as well as where, to work will therefore enhance efficiency and overall performance, albeit there may be some practical issues to be resolved for those working closely in teams. Freedom over when to work also supports better work-life balance, allowing personal commitments to be fit into the day. Agile working practices help meet human needs and go hand in hand with the *Landscaped Office* design and layout.

The post-pandemic workplace will be different, but maybe not so different for those who already practiced agile working. Fundamentally, post-pandemic "hybrid working" is agile working with reliable and seamless technology championed by management. There will also be more emphasis on wellbeing and reducing health risks along with a more balanced approach to working in and out of the office whilst maintaining motivation and performance. This shift is long overdue, regardless of the pandemic, but nevertheless presents an opportunity to rethink the office. Lower densities, a richer range of work-settings, more screens and zones, improved environmental conditions, and biophilic design will all help create a healthier workplace.

Clearly, costs must be managed in any organisation, but the trick with planning and designing the office is to not be constrained by cost and space at the start, before even understanding the requirements and commencing the design. If cost control is the priority, then consider agile working as part of the solution rather than densification and space reduction. For example, if there is concern about whether the organisation will fit in the workspace, then use say 15 m$^2$ per desk rather than the typical 10 m$^2$ per desk in the UK and assume that 60% of the desks are required to accommodate the assigned building population, equivalent to 9 m$^2$ per person. Nonetheless, design should commence with the intention of maximising the wellbeing and performance of the occupants, tapping into their needs highlighted through documented psychological research and through extensive consultation. In essence, the workplace design should be human-centric and evidence-based rather than be solely driven by cost or how cool it looks.

Figure 10.1 illustrates the physical components of the *Landscaped Office*. Such a working environment is underpinned by enabling technology, covered in detail by others such as John Eary (2018). The success of the *Landscaped Office* is also dependent upon a corresponding shift in organisational culture and management style. Like any new workplace, the transition to the *Landscaped Office* and even more so to the *Agile Landscaped Office* (ALO) requires planning, consultation and change management.

## Implementing the *Landscaped Office*

While I have illustrated the many different work-settings of the *Landscaped Office*, and their layout, I have not specified the numbers required. Determining the appropriate number of desks, work-settings and corresponding space is a challenge and depends upon personal requirements, typical work activities and the time typically spent in the office. These factors can only be understood through a thorough briefing process, including consultation with a cross-section of the office occupants and research, such as observation studies. The briefing may be conducted by the in-house design team or by specialist workplace consultants.

Moving from a traditional workplace layout to the *Landscaped Office* will most certainly require a change management programme. Such a programme prepares and assists the occupants in transitioning to their new workplace. At minimum, change programmes cover the "six Ws": <u>w</u>hat, <u>w</u>hy, <u>w</u>hen, <u>w</u>here, <u>w</u>ho and ho<u>w</u>. The change programme will include engagement with various stakeholders through interviews, surveys, workshops, townhall meetings, bulletin boards, newsletters, intranet sites, furniture mock-up and "beauty parades", site tours, pilot studies, etc. For a comprehensive insight into change management, I recommend the books of my fellow workplace advisors: Neil Usher's (2020) *Elemental Change* and Jennifer Bryan's (2021) *Leading People in Change: A Practical Guide*. Alternatively, for a quick high-level approach, see my own *Seven Cs of Change* (Oseland, 2013c), based on Kotter's (1996) eight-step process and other change models.

The successful move to any new workplace concept requires strong leadership, with senior stakeholders bought into it and leading by example. It also requires a commitment by the organisation to the finances and resources required to implement it. A successful workplace is dependent upon its use, and the behaviour of its occupants as well as its design. Therefore, sessions on how to use the space are required including the development of any applicable office protocols. Agile working environments require training for team leaders on how to manage remote teams and wider training in how to effectively work from home and how to use new technology applications.

## Beyond this book

The primary intention of this book is to share with you the vast psychological literature relevant to workplace design by distilling it into an easily digested text. Over my years of workplace consulting, I increasingly saw offices designed from the wrong perspective, transforming them into a *Workplace Zoo*, more akin to those early desolate zoos with rows of cages rather than the modern well-designed animal enclosures. Basic psychophysical requirements are often overlooked, and innate human needs are rarely considered, resulting in high-density unattractive and, from a viral perspective, unsafe, unhealthy workplaces.

The relatively recent shift to designing with wellbeing in mind, rather than cost alone, and the uptake of biophilic design demonstrates that things are gradually improving. The post-pandemic era provides the ideal opportunity to rethink our office space and design primarily with people in mind, catering for their psychological and physiological needs. Hopefully, the lower utilisation of offices going forward means that design errors of the past can be rectified, and open plan design reviewed and reclaimed as good design. The advent of hybrid (office and home) working, and adoption of agile working environments, means that the balance of space can be addressed – providing the right spaces to facilitate the wellbeing and performance of all the workplace inhabitants.

I have presented you with my revised version of the *Landscaped Office*, and hopefully, I have provided the relevant information, and inspiration, to inform your own workplace layouts and designs. Your next step begins with a robust business case, comprehensive occupant consultation, diligent design briefing and a well-structured change management programme.

# References

Abraham, Y. (1998). The man behind the cubicle. *Metropolis, November.*

AECOM (2014). The next generation occupier. Cited in *Future Workstyles and Future Workplaces in the City of London* (2015). London: City of London Corporation and The City Property Association.

AIS (2011). *A Guide to Office Acoustics.* Solihull: Association of Interior Specialists.

Alker, J. (2014). *Health, Wellbeing & Productivity in Offices: The Next Chapter for Green Buildings.* London: World Green Building Council (WGBC).

Allen, J.G., MacNaughton, P., Satish, U., Santanam, S., Vallarino, J. & Spengler, J.D. (2016). Associations of cognitive function scores with carbon dioxide, ventilation, and volatile organic compound exposures in office workers: A controlled exposure study of green and conventional office environments. *Environmental Health Perspectives, 124(6),* 805–812.

Allen, T., Bell, A., Graham, R., Hardy, B. & Swaffer, F. (2004). *Working Without Walls: An Insight into the Transforming Government Workplace.* London: Office of Government Commerce.

Allen, T.J. & Fusfeld, A. (1975). Research laboratory architecture and the structuring of communications. *R&D Management, 5(2),* 153–163.

Allport, G.W. & Odbert, H.S. (1936). Trait-names: A psycho-lexical study. *Psychological Monographs, 47(1),* i–171.

Allsopp, P. (2010). What is agile working? *Personnel Today.* Retrieved from https://www.personneltoday.com/hr/articles/what-is-agile-working/.

Altman, I. (1975). *The Environment and Social Behavior.* Monterey: Brooks/Cole.

ASHRAE (2017). *Standard 55 – Thermal Environmental Conditions for Human Occupancy.* Atlanta: American Society of Heating, Refrigerating and Air-Conditioning Engineers.

ASHRAE (2019). *Standard 62.1–2019 – Ventilation for Acceptable Indoor Air Quality.* Atlanta: American Society of Heating, Refrigerating and Air-Conditioning Engineers.

Atchley, R.A., Strayer, D.L. & Atchley, P. (2012). Creativity in the wild: Improving creative reasoning through immersion in natural settings. *PLoS ONE, 7(12),* e51474.

Attema, J.E., Fowell, S.J., Macko, M.J. & Neilson, W.C. (2018). *The Financial Case for High Performance Buildings: Quantifying the Bottom Line of Improved Productivity, Retention and Wellness*. San Francisco: Stok LLC.

Barker, R.B. (1968). *Ecological Psychology: Concepts and Methods for Studying the Environment of Human Behavior.* Stanford: Stanford University Press.

Barnaby, J.F. (1980). Lighting for productive gains. *Lighting Design Application, 10,* 20–28.

Barney, T. & Hodsman, P. (2020). *A Sensory Processing Approach to Office Acoustic Design.* Report published by Saint-Gobain Ecophon, Tadley.

BCO (2014). *BCO Guide to Specification 2014.* London: British Council for Offices.

BCO (2018). *Office Occupancy: Density and Utilisation.* London: British Council for Offices.

BCO (2020). Majority of workers plan a return to the office, but home working here to stay. *BCO News, October.* Retrieved from https://www.bco.org.uk/News/News46982.aspx.

Bedford, M. Harris, R., King, A. & Hawkeswood, A. (2013). *BCO Office Occupation Density Study 2013.* London: British Council for Offices.

Bernstein, E.S. & Turban, S. (2018). The impact of the 'open' workspace on human collaboration. *Philosophical Transactions B, 373(1753),* 20170239.

BIFM (2016). *The Workplace Advantage, The Stoddart Review.* Bishop's Stortford: British Institute of Facilities Management.

Borisuit, A., Linhart, F., Scartezzini, J.-L. & Münch, M. (2015). Effects of realistic office daylighting and electric lighting conditions on visual comfort, alertness and mood. *Lighting Research & Technology, 47(2),* 192–209.

Bowers, A.R., Meek, C. & Stewart, N. (2001). Illumination and reading performance in age related macular degeneration. *Clinical and Experimental Optometry, 84(3),* 139–147.

Boyce, P.R., Veitch, J.A., Newsham, G.R., Jones, C.C., Heerwagen, J., Myer, M. & Hunter, C.M. (2006). Occupant use of switching and dimming controls in offices. *Lighting Research & Technology, 38(4),* 358–376.

Brem, A. (2019). The biggest problem with open plan offices is how they are used. *Workplace Insight, 1 July.* Retrieved from https://workplaceinsight.net/the-biggest-problem-with-open-plan-offices-is-how-they-are-used/.

Briggs, K. & Myers, I. (1987). *Myers-Briggs Type Indicator Form G.* Palo Alto: Consulting Psychologist Press.

Briggs, S.P., Copeland, S. & Haynes, D. (2002). Accountants for the 21st Century, where are you?: A five-year study of accounting students' personality preferences. *Critical Perspectives on Accounting, 18(5),* 511–537.

Broadbent, D. (1958). *Perception and Communication.* Oxford: Pergamon Press.

Broady, M. (1966). Social theory in architectural design. *Arena: The Architectural Association Journal, 81,* 149–154.

Browning, B. et al. (2012). *The Economics of Biophilia: Why Designing with Nature in Mind Makes Financial Sense.* New York: Terrapin Bright Green.

Browning, W., Ryan, C. & Clancy, J. (2014). *14 Patterns of Biophilic Design: Improving Health & Well-Being in the Built Environment.* New York: Terrapin Bright Green.

Bryan, J. (2021). *Leading People in Change: A Practical Guide*. London: Hero Publishing.

BSI (2005). *BS 7000–6:2005: Design Management Systems - Managing Inclusive Design*. London: British Standards Institute.

BSI (2007). *BS EN 15251:2007: Indoor Environmental Input Parameters for Design and Assessment of Energy Performance of Buildings Addressing Indoor Air Quality, Thermal Environment, Lighting and Acoustics*. London: British Standards Institute.

BSI (2014). *BS 8233:2014: Guidance on Sound Insulation and Noise Reduction for Buildings*. London: British Standards Institute.

Buckley, J.P., Mellor, D.D., Morris, M. & Joseph, F. (2014). Standing-based office work shows encouraging signs of attenuating post-prandial glycaemic excursion. *Occupation and Environmental Medicine, 71(2)*, 109–111.

Burns, C.M. (2018). Are open plan workspaces truly evil or is this just fake news? *Workplace Insight, 21 December*.

Burt, K., Oseland, N.A., Marks, K. & Greenberg, B. (2010). *Making Flexible Working Work*. London: British Council for Offices.

CABE & BCO (2006). *The Impact of Office Design on Business Performance*. London: The Commission for Architecture and the Built Environment, and British Council for Offices.

Cain, S. (2012). *Quiet: The Power of Introverts in a World That Can't Stop Talking*. New York: Penguin Random House.

Calhoun, J.B. (1962). Population density and social pathology. *Scientific American, 206(3)*, 139–148.

Cambridge University Press (2021). *Cambridge Dictionary*. Retrieved from https://dictionary.cambridge.org/ dictionary/english/remote-working

Card, J. (2014). Tips for boosting productivity with good office design. *The Guardian, 23 January*. Retrieved from https://www.theguardian.com/small-business-network/2014 /jan/23/productivity-office-design.

Cattell, R.B. (1947). Confirmation and clarification of primary personality factors. *Psychometrika, 12(3)*, 197–220.

CBRE (2015). *Space Utilisation: The Next Frontier – How Asian Market Conditions are Driving Utilisation Harder and Faster*. CBRE Workplace Strategy Research Report.

CIBSE (2015). *Environmental Design: CIBSE Guide A (8th Edition)*. London: Chartered Institution of Building Services.

CIPD (2104). *HR: Getting Smart about Agile Working, Research Report, November 2014*. London: Chartered Institute of Personnel and Development.

Claudio, L. (2011). Planting healthier indoor air. *Environmental Health Perspectives, 119(10)*, A426–A427.

Costa, P.T. & McCrae, R.R. (1992). *Revised NEO Personality Inventory (NEO-PI-R) and NEO Five-Factor Inventory (NEO-FFI) Manual*. Odessa: Psychological Assessment Resources.

Dart, R. (1925). Australopithecus Africanus: The man-ape of South Africa. *Nature, 115 (2884)*, 195–199.

Davis, M.C., Leach, D.J. & Clegg, C.W. (2011). The physical environment of the office: Contemporary and emerging issues. Chapter 6 in G.P. Hodgkinson & J.K.

Ford (Eds.) *International Review of Industrial and Organizational Psychology, 26,* 193–237. Chichester: Wiley-Blackwell.

Dawkins, R. (1976). *The Selfish Gene.* Oxford: Oxford University Press.

de Vries, A, Souman, J.L., de Ruyter, B., Heynderickx, I. & de Kort, Y.A.W. (2018). Lighting up the office: The effect of wall luminance on room appraisal, office workers' performance, and subjective alertness. *Building and Environment, 142,* 534–543.

Defra (2007). *Sustainable Development Indicators in Your Pocket 2007: An Update of the UK Government Strategy Indicators.* London: Department for Environment, Food and Rural Affairs.

Defra (2012). *Secretary of State's Standards of Modern Zoo Practice.* London: Department for Environment, Food and Rural Affairs.

Delos Living (2015). *WELL Building Standard v1.0.* WELL Building Institute. Retrieved from https://standard.wellcertified.com/sites/standard.wellcertified.com. v3/ files/The%20WELL%20Building%20Standard%20September%202015.pdf.

Despenic, M., Chraibi, S., Lashina, T. & Rosemann, A. (2017). Lighting preference profiles of users in an open office environment. *Building and Environment, 116,* 89–107.

DETR (2000). *Secretary of State's Standards of Modern Zoo Practice.* London: Department of the Environment, Transport and the Regions.

DETR (2010). *Welfare of Animals During Transport Council Regulation (EC) No 1/2005 on the Protection of Animals During Transport and Related Operations and The Welfare of Animals (Transport) (England) Order 2006.* London: Department of the Environment, Transport and the Regions.

Dubner, S.J. (2018). Yes, the open office is terrible – But it doesn't have to be. *Freakonomics, Episode 358, 14 November.*

Duffy, F. (1966). *Office Landscaping: A New Approach to Office Planning.* Bingley: Anbar Abstracts MCB University Press Ltd.

Duffy, F. (1997). *The New Office.* London: Conran Octopus Limited.

Dunbar, R.I.M. (1992). Neocortex size as a constraint on group size in primates. *Journal of Human Evolution, 22(6),* 469–493.

Dunn, W. (1997). The impact of sensory processing abilities on the daily lives of young children and their families: A conceptual model. *Infants & Young Children, 9(4),* 23–35.

Durie, B. (2005). Senses special: Doors of perception. *New Scientist, 2484,* 34.

Eary, J. (2018). *Agile Working and the Digital Workspace: Best Practices for Designing and Implementing.* New York: Business Expert Press.

EAZA (2013). *The Modern Zoo: Foundations for Management and Development.* Amsterdam: European Association of Zoos and Aquaria.

Eley, J. & Marmot, A. (1995). *Understanding Offices: What Every Manager Needs to Know about Office Buildings.* London: Penguin Business.

Erikson, C. & Küller, R. (1983). Non-visual effects of office lighting. *CIE 20th Session, Amsterdam, Volume 1.* Commission Internationale de L'Eclairage, Vienna.

Eysenck, H.J. (1947). *Dimensions of Personality.* London: Kegan Paul.

Eysenck, H.J. & Eysenck, S.B.G. (1975). *Manual of the Eysenck Personality Questionnaire (Junior and Adult).* Kent: Hodder & Stoughton.

Fanger, P.O. (1970). *Thermal Comfort. Analysis and Applications in Environmental Engineering.* Copenhagen: Danish Technical Press.

Fechner, G.T. (1860). *Elemente der Psychophysik.* Leipzig: Breitkopf und Härtel.

Ferreira, E.J., Erasmus, A.W. & Groenewald, D. (2003). *Administrative Management.* Lansdowne: Juta and Company Ltd.

Freud, S. (1901). *Zur Psychopathologie des Alltagslebens.* Published in English as *The Psychopathology of Everyday Life* (2002). London: Penguin Classics.

Fromm, E. (1973). *The Anatomy of Human Destructiveness.* New York: Holt, Rinehart and Winston.

Gates, B. (2005). *Digital Workstyle: The New World of Work.* Speech at Microsoft CEO Summit. Retrieved from http://www.finfacts.ie/irelandbusinessnews/publish/ article_10001879.shtml.

Gensler (2013). *2013 U.S. Workplace Survey: Key Findings.* San Francisco: Gensler.

Gensler (2019). New data changes the conversation on the open office. San Francisco: Gensler Press Release. Retrieved from https://www.gensler.com/news/press-releases/new-data-changes-the-conversation-on-the-open-office.

Gerald Eve (2001). *Overcrowded, Underutilised or Just Right?* London: Gerald Eve.

Gerba, C.P. (2002). Germs in the workplace. Unpublished study by University of Arizona.

Global Workplace Analytics (2020). Latest work-at-home/telecommuting/mobile work/ remote work statistics. *Global Workplace Analytics, March.* Retrieved from https://globalworkplaceanalytics.com/ telecommuting-statistics..

Goldsmith, S. (1963). *Designing for the Disabled: A Manual of Technical Information.* London: Royal Institute of British Architects.

Hall, E.T. (1963). A system for the notation of proxemic behaviour. *American Anthropologist, 65,* 1003–1026.

Hamilton, D. K. & Watkins, D. (2009). What is evidence-based design. Chapter 1 in Hamilton, D.K. and Watkins, D.H. (eds) *Evidence Based Design for Multiple Building Types.* New York: John Wiley and Sons.

Harris, R. (2021). *London's Global Office Economy: From Clerical Factory to Digital Hub.* London: Routledge.

Harris, R., Bedford, M., Gillen, N., Jack, F. & Whitehead, C. (2018). *Office Occupancy: Density and Utilisation.* London: British Council for Offices.

Harris, R. & Hawkeswood, A. (2016). *The Proportion of Underlying Business Costs Accounted for by Real Estate.* London: British Council for Offices.

Hediger, H. (1942). *Wildtiere in Gefangenschaft.* Published in English as *Wild Animals in Captivity.* Translated by G. Sircom (1950). London: Butterworth.

Heerwagen, J. (2008). *Psychosocial Value of Space.* J.H. Seattle: Heerwagen & Associates, Inc.

Heisenberg, W. (1927). Über den anschaulichen Inhalt der quantentheoretischen Kinematik und Mechanik. *Zeitschrift für Physik, 43(3–4),* 172–198.

Herman Miller (2013). *Living Office: The Origins of Herman Miller's Modes of Work.* Michigan: Herman Miller Inc.

Herman, C.M. (1969). The impact of disease on wildlife populations. *BioScience, 19(4),* 321–330.

Herzberg, F. (1959). *The Motivation to Work.* New York: John Wiley & Sons.

Heschong, L., Saxena, M., Wright, R., Okura, S. & Aumann, D. (2004). Offices, windows and daylight: Call center worker performance. *Proceedings from ACEEE Summer Studies on Energy Efficiency in Buildings, Panel 7*, 98–110. European Council for an Energy Efficient Economy.

Hiroshi, S., Kazuhiro, T., Hirotake, I., Fumiaki, O., Masaaki, T. & Yoshikawa, H. (2006). A study on an environment control method to improve productivity of office worker. *Proceeding of Sustainable Energy and the Environment (SEE) International Conference*. Bangkok.

Hodsman, P. & Oseland, N.A. (2021). *Design Guidance on Reducing Office Noise: A Psychoacoustic Approach*. Tadley: Saint-Gobain Ecophon.

Hongisto, V. Haapakangas, A., Varjo, J., Helenius, R. & Koskela, H. (2016). Refurbishment of an open-plan office – Environmental and job satisfaction. *Journal of Environmental Psychology, 45*, 176–191.

Hongisto, V., Varjo, J., Oliva, D., Haapakangas, A. & Benway, E. (2017). Perception of water-based masking sounds-long-term experiment in an open-plan office. *Frontiers in Psychology, 8*, 1177.

HRZone (2017). New ways of working: Transforming the way you work. *HRZone*. Retrieved from https://www.hrzone.com/community/blogs/karl-elliot/new-ways-of-working-transforming-the-way-you-work.

ISO (2005). *ISO 7730:2005: Ergonomics of the Thermal Environment — Analytical Determination and Interpretation of Thermal Comfort Using Calculation of The PMV and PPD Indices and Local Thermal Comfort Criteria*. Geneva: International Organization for Standardization.

ISO (2019). *ISO 9241-210:2019(en) Ergonomics of Human-System Interaction – Part 210: Human-Centred Design for Interactive Systems*. Geneva: International Organization for Standardization.

ISO (2020). *ISO/FDIS 22955 Acoustics – Acoustic Quality of Open Office Spaces*. Geneva: International Organization for Standardization. Under development.

Jaffe, E. (2015). Morning people vs. night owls: 9 insights backed by science. *Fast Company, 19 May*.

JEP (2021). *Journal of Environmental Psychology: Author Information Pack*. Amsterdam: Elsevier. Retrieved from https://www.elsevier.com/wps/find/journaldescription.cws_ home/622872?generatepdf=true.

John, O.P., Robins, R.W. & Pervin, L.A. (2011). *Handbook of Personality: Theory and Research (3rd Edition)*. New York: The Guilford Press.

Johnson, A. (2010). *Shedworking: The Alternative Workplace Revolution*. London: Frances Lincoln Limited.

Jung, C.G. (1912). *Wandlungen und Symbole der Libido*. Published in English as *Psychology of the Unconscious*. Eastford: Martino Fine Books (2016).

Kane, C. (2020). *Where Is My Office: Reimagining the Office for the 21st Century*. London: Bloomsbury Business.

Katjár, L. & Herczeg, L. (2012). Influence of carbon-dioxide concentration on human well-being and intensity of mental work. *Quarterly Journal of the Hungarian Meteorological Service, 116*, 145–169.

Kellert, S. (2015). What is and is not biophilic design? *Metropolis Magazine, October*.

Kellert, S. & Calabrese, E. (2015). *The Practice of Biophilic Design*. Retrieved from www.biophilic-design.com.

Kellert, S.R., Heerwagen, J. & Mador, M. (2008). *Biophilic Design: The Theory, Science and Practice of Bringing Buildings to Life*. Hoboken: John Wiley & Sons.

Kim, J. & de Dear, R. (2013). Workspace satisfaction: The privacy-communication trade-off in open-plan offices. *Journal of Environmental Psychology, 36*, 18–26.

Knight, C. & Haslam, S.A. (2010). The relative merits of lean, enriched, and empowered offices: An experimental examination of the impact of workspace management strategies on well-being and productivity. *Journal of Experimental Psychology, 16(2)*, 158–172.

Kotter, J.P. (1996). *Leading Change*. Boston: Harvard Business School Press.

Kroner, W.J., Stark-Martin, J.A. & Willemain, T. (1992). *Using Advanced Office Technology to Increase Productivity: The Impact of Environmentally Responsive Workstations (ERWs). On Productivity and Worker Attitude*. Rensselaer: The Center for Architectural Research.

Lan, L., Wargocki, P. & Lian, Z. (2011). Quantitative measurement of productivity loss due to thermal discomfort. *Energy and Buildings 43(5)*, 1057–1062.

Lan, L., Wargocki, P., Wyon, D.P. & Lian, Z. (2011). Effects of thermal discomfort in an office on perceived air quality, SBS symptoms, physiological responses, and human performance. *Indoor Air, 21(5)*, 376–390.

Lan, L., Xia, L., Hejjo, R., Wyon, D.P. & Wargocki, P. (2020). Perceived air quality and cognitive performance decrease at moderately raised indoor temperatures even when clothed for comfort. *Indoor Air, 30(5)*, 841–859.

Le Corbusier (1927). *Vers Une Architecture (Towards a New Architecture)*. Translated by Etchells, F. London: J. Rodker.

Leaman, A. & Bordass, B. (2010). Productivity in buildings: The 'killer' variables. *Building Research and Information January 1(1)*, 4–19.

Lee, J.-H., Moon, J.W. & Kim, S. (2014). Analysis of occupants' visual perception to refine indoor lighting environment for office tasks. *Energies 7(7)*, 4116–4139.

Leesman (2019a). *The World's Best Workplaces 2018: Lessons from the Leaders in Employee Experience*. London: Leesman.

Leesman (2019b). *The Workplace Experience Revolution: Part 2: Do New Workplaces Work*. London: Leesman.

Levin, H. (1992). *Workshop on Productivity and the Indoor Environment: Preface to Workshop Proceedings*. Baltimore: ASHRAE.

Lewin, K. (1936). *Principles of Topological Psychology*. New York: McGraw-Hill.

Lister, K. (2014). *What's Good for People? Moving from Wellness to Well-Being*. New York: Knoll.

Loftness, V.E. et al. (2007). Health, productivity and the triple bottom line. *Presentation to the Center for Building Performance and Diagnostics, an NSF/IUCRC and ABSIC at Carnegie Mellon*. Retrieved from https://www.yumpu.com/en/document/view/8032471/ health-productivity-and-the-triple-bottom-line-carnegie-mellon-.

Lohr, S. (1997). Rethinking privacy vs. teamwork in today's workplace. *The New York Times, 11 August*. Retrieved from https://www.nytimes.com/1997/08/11/business/ rethinking-privacy-vs-teamwork-in-today-s-workplace.html.

Lombard, A. (2007). *Sensory Intelligence: Why It Matters More Than Both IQ and EQ*. Cape Town: Metz Press.

Loneliness Lab (2020). *We're Designing Loneliness out of the Workplace*. London: Loneliness Lab.

Mace, R. (1985). Universal design: Barrier-free environments for everyone. *Designers West, 33(1)*, 147–152.

Mace, R. et al. (1997). *The Principles of Universal Design, Version 2.0*. North Carolina State University, The Center for Universal Design. Retrieved from https:// projects. ncsu.edu/ design/cud/pubs_p/docs/poster.pdf.

MacLean, P.D. (1967). The brain in relation to empathy and medical education. *Journal of Nervous and Mental Disease, 144*, 374–382.

MacLean, P.D. (1990). *The Triune Brain in Evolution: Role in Paleocerebral Functions*. New York: Plenum Press.

Malmberg, T. (1980). *Human Territoriality: Survey of Behavioral Territories in Man with Preliminary Analysis and Discussion of Meaning*. New York: Mouton.

Marsh, M. & Mueller, K. (2017). Multisensory design: The empathy-based approach to workplace wellness. *WorkDesign Magazine*. Retrieved from https://www.workdesign. com/2017/04/multisensory-design-empathy-based-approach-workplace-wellness.

Maslin, S. (2009). Inclusive design. *Construction Week Online, 28 December*. Retrieved from http://www.constructionweekonline.com/article-7264-inclusive-design/.

Maslow, A.H. (1943). A theory of human motivation. *Psychological Review 50*, 370–396.

Matthews, V. (2006). Scents and sensibilities: Is the smell of cinnamon or sandalwood really capable of lightening the mood and enhancing productivity? *The Guardian*. https://www.theguardian.com/money/2006/may/08/careers.theguardian5

Maula, H., Hongisto, V., Naatula, V., Haapakangas, A. & Koskela, H. (2017). The effect of low ventilation rate with elevated bioeffluent concentration on work performance, perceived indoor air quality, and health symptoms. *Indoor Air, 27(6)*, 1141–1153.

Mayo, E. (1933). Volume VI: The human problems of an industrial civilization. In K. Thompson (Ed.) *The Early Sociology of Management and Organizations*. 171–171, New York: Macmillan.

Megginson, L.C. (1963). Lessons from Europe for American Business. *Southwestern Social Science Quarterly, 44(1)*, 3–13.

Milind, P., Jyoti, M. & Sushila, K. (2013). Life style related health hazards. *International Research Journal of Pharmacy, 4(11)*, 1–5.

Mixson, E. (2019). In defence of open plan office design. *Workplace Insight, 6 June*. Retrieved from https://workplaceinsight.net/in-defence-of-open-plan-office-design/.

Monice Malnar, J. & Vodvarka, F. (2004). *Sensory Design*. Minnesota: University of Minnesota Press.

Moore, T., Carter, D. & Slater, A. (2003). Long-term patterns of use of occupant controlled office lighting. *Lighting Research & Technology, 35(1)*, 43–57.

Morris, D. (1967). *The Naked Ape: A Zoologist's Study of the Human Animal*. London: Jonathan Cape.

Morris, D. (1969). *The Human Zoo*. London: Jonathan Cape.

Myerson, J., Bichard, J.-A. & Erlich, A. (2010). *New Demographics, New Workspace: Office Design for the Changing Workforce*. Aldershot: Gower Publishing.

National Bureau of Standards (1964). Study of federal office buildings. In Office of the Sate Architect, California *Building Value, Energy Design Guidelines for Sate Buildings*. Cited by P. Wargocki. How indoor climate affects productivity in office, schools and similar buildings. Retrieved from https://www.innobyg.dk/media/52235/pawel%20 wargocki%20I%C3%A5st.pdf.

NEMA (1989). *Lighting and Human Performance: A Review*. Washington: National Electrical Manufactures Association.

Newsham, G., Veitch, J., Arsenault, C. & Duval, C. (2004). *Effect of Dimming Control on Office Worker Satisfaction and Performance*. Ottawa: Institute for Research in Construction/National Research Council Canada.

Ohly, H. White, M.P., Wheeler, B.W. Bethel, A., Ukoumunne, O.C., Nikolaou, V. & Garside, R. (2016). Attention restoration theory: A systematic review of the attention restoration potential of exposure to natural environments. *Journal of Toxicology and Environmental Health, Part B, 19(7)*, 305–343.

Oldman, T. (2016). *Figures confirm stagnant labour productivity levels*. Press release by Megenta Associates on behalf of Leesman. Retrieved from https://pressreleases.responsesource.com/news/91046/figures-confirm-stagnant-labour-productivity-levels/.

Oommen, V.G., Knowles, M. & Zhao, I. (2008). Should health service managers embrace open plan work environments? A review. *Asia Pacific Journal of Health Management, 3(2)*, 37–43.

Orians, G.H. & Heerwagen, J.H. (1992). Evolved responses to landscapes. In J.H. Barkow, L. Cosmides & J. Tooby (Eds.), *The Adapted Mind: Evolutionary Psychology and the Generation of Culture*, 555–579. Oxford: Oxford University Press.

Oseland, N.A. (1995). Predicted and reported thermal sensation in climate chambers, offices and homes. *Energy & Buildings, 23*, 105–115.

Oseland, N.A. (1999). *Environmental Factors Affecting Office Worker Performance: Review of Evidence, Technical Memorandum TM24: 1999*. London: The Chartered Institution of Building Services Engineers.

Oseland, N.A. (2012). *The Psychology of Collaboration Space*. London: Herman Miller.

Oseland, N.A. (2013a). The bigger the better: Design trends in law firms. *Facilities Management, September*, 13–15.

Oseland, N.A. (2013b). *Personality and Preferences for Interaction, WPU-OP-03*. Workplace Unlimited Occasional Paper.

Oseland, N.A. (2013c). Seven Cs of change. *Workplace Unlimited Blogspot*. Retrieved from http://workplaceunlimited.blogspot.com/2013/03/seven-cs-of-change.html.

Oseland, N.A. & Bartlett, P. (1999). *Improving Office Productivity: A Guide for Business and Facilities Managers*. Harlow: Longman.

Oseland, N.A. & Burton, A. (2012). Quantifying the impact of environmental conditions on worker performance for inputting to a business case. *Journal of Building Survey, Appraisal and Valuation, 1(2)*, 151–165.

Oseland, N.A. & Catchlove, M. (2020). Personal office preferences. Proceedings of *Transdisciplinary Workplace Research Conference, TWR 2020*. Frankfurt.

Oseland, N.A. & Donald, I. (1993). The evaluation of space in homes: A facet study. *Journal of Environmental Psychology, 13(3)*, 251–261.

Oseland, N.A., Fargus, R. Brown, D.K. Charnier, A.D. & Leaman, A.J. (1997). DUCOZT, a prototype system for democratic user control of zonal temperature in airconditioned offices. *CIBSE 97 Conference: Quality for People*, 1–15.

Oseland, N.A. & Hodsman, P. (2017). Psychoacoustics: Resolving noise distractions in the workplace. Chapter 4 in A. Hedge (Ed.) *Ergonomics Design for Healthy and Productive Workplaces*. 73–101, Abingdon: Taylor & Francis.

Oseland, N.A. & Hodsman, P. (2020). The response to noise distraction by different personality types: An extended psychoacoustics study. *Corporate Real Estate Journal, 9(3)*, 215–233.

Oseland, N.A., Humphreys, M.A., Nicol, J.F., Baker, N.V. & Parsons, K.C. (1998). *Building Design and Management for Thermal Comfort (CR 203/98)*. Client report prepared for CIBSE. Watford: Building Research Establishment.

Oseland, N.A., Tucker, M. & Wilson, H. (2021). *In pursuit of the 'Holy Grail': Determining the Return on Workplace Investment*. Bishop's Stortford: Institute of Workplace and Facilities Management. In press.

Osmond, H. (1957). Function as the basis of psychiatric ward design. *American Psychiatric Association: Mental Hospitals (Architectural Supplement), 8*, 23–29.

Palmer, S.E. & Schloss, K.B. (2010). An ecological valence theory of human color preference. *Proceedings of the National Academy of Sciences of the United States of America, 107(19)*, 8877–8882.

Park, B.J., Tsunetsugu, Y., Kasetani, T., Kagawa, T. & Miyazaki, Y. (2010). The physiological effects of Shinrin-yoku (taking in the forest atmosphere or forest bathing): Evidence from field experiments in 24 forests across Japan. *Environmental Health and Preventive Medicine, 15(1)*, 18–26.

Park, N.-K. & Farr, C.A. (2007). Retail store lighting for elderly consumers: An experimental approach. *Journal of Family and Consumer Sciences, 35(4)*, 316–337.

Parker, C. & Oseland, N.A. (2019). *Creating the Perfect Meeting Environment*. Montvale: Sharp. Retrieved from https://business.sharpusa.com/White-Papers/Creating-the-Perfect-Meeting-Environment.

Pejtersen, J.H., Feveile, H., Christensen, K.B. & Burr, H. (2011). Sickness absence associated with shared and open-plan offices - A national cross-sectional questionnaire survey. *Scandinavian Journal of Work Environmental Health, 37(5)*, 376–382.

Pell, N. (2021). The free-range office. Personal email correspondence.

Pogue McLaurin, J. (2018). The open office isn't dead. *Gensler Research & Insight*. San Francisco: Gensler. Retrieved from https://www.gensler.com/research-insight/blog/the-open-office-isnt-dead.

Pollard, E.L. & Lee, P.D. (2003). Child well-being: A systematic review of the literature. *Social Indicators Research, 61(1)*, 59–78.

Preiser, W.F.E., Rabinowitz, H.R. & White, E.T. (1988). *Post Occupancy Evaluation*. New York: Van Nostrand Reinhold.

Propst, R. (1998). Cited by Nikil Saval (2014) *Cubed: A Secret History of the Workplace*. New York: Doubleday & Co.

PWC (2021). It's time to reimagine where and how work will get done: PwC's US remote work survey. *PWC Research and Insights, January*. Retrieved from https://www.pwc.com/us/en/library/covid-19/us-remote-work-survey.html.

Raisbeck, K. (2003). *Productivity in the Workplace, MBA Dissertation*. Henley: Henley Management College.

Risner, N. (2003). *"It's a Zoo Around Here": The New Rules for Better Communication*. Arkley: Limitless Publications.

Risner, N. (2020). *Zoo Keeper Rules for the Office: Tame the Beasts, Build a Team and Thrive*. Arkley: Limitless Publications.

Ritter, S.M. (2012). *Creativity: Understanding and Enhancing Creative Thinking*. PhD Thesis. Radboud Universiteit Nijmegen, Nijmegen.

Roethlisberger, F.J. & Dickson, W.J. (1939). *Management and the Worker*. Cambridge: Harvard University Press.

Rohles, F.H. & Wells, W.V. (1976). Interior design, comfort and thermal sensitivity. *Journal of Interior Design, Education, and Research, 2(2)*, 36–44.

Rothe, P. (2017a). *The Open Plan Witch Hunt*. London: Leesman. Retrieved from https://www.leesmanindex.com/the-open-plan-witch-hunt/.

Rothe, P. (2017b). *The Rise and Rise of Activity Based Working*. London: Leesman.

Rotter, J.B. (1966). Generalized expectancies of internal versus external control of reinforcements. *Psychological Monographs, 80(1)*, 609.

Sadalla, E.K., & Oxley, D. (1984). The perception of room size: The rectangularity illusion. *Environment and Behavior, 16(3)*, 394–405.

Sargent, K. et al. (2019). *Designing a Neurodiverse Workplace*. New York: HOK Group Inc. Retrieved from https://www.hok.com/ideas/publications/hok-designing-a-neurodiverse-workplace.

Satish, U., Mendell, M.J., Shekhar, K., Hotchi, T., Sullivan, D., Streufert, S. & Fisk, W.J. (2012). Is $CO_2$ an indoor pollutant? Direct effects of low-to-moderate $CO_2$ concentrations on human decision-making performance. *Environmental Health Perspectives, 120(12)*, 1671–1677.

Schaubhut, N.A. & Thompson, R.C. (2008). *MBTI Type Tables for Occupations*. Mountain View: CPP Inc.

Scheiberg, S.L. (1990). Emotions on display: The personal decoration of work space. *American Behavioral Scientist, 33(3)*, 330–338.

Seppänen, O.A. & Fisk, W. (2006). Some quantitative relations between indoor environmental quality and work performance or health. *HVAC&R Research, 12(4)*, 957–973.

Seppänen, O., Fisk, W. & Lei, Q. (2006). Effect of temperature on task performance in office environments. *Proceedings of Cold Climate HVAC Conference*. Moscow.

Sidders, J. (2019). WeWork squeezes people into just half the space of most offices. *BloombergQuint*. Retrieved from https://www.bloombergquint.com/onweb/tight-squeeze-wework-jams-more-folks-into-its-space-than-others.

Sommer, R. (1967). Sociofugal space. *American Journal of Sociology, 27(6)*, 654–660.

Sorokowska, A. et al. (2017). Preferred interpersonal distances: A global comparison. *Journal of Cross-Cultural Psychology, 48(4)*, 577–592.

Souza, E. (2019). How lighting affects mood. *ArchDaily, 12 August*.

Sowa, K. (2020). Creation of the indoor environment in office buildings. *ASHRAE Journal, 62(7)*, 64–68.

Stangl, K. (2017). *The Workplace Zoo*. New Mexico: Mercury HeartLink.

Steelcase (2012). Culture code: Leveraging the workplace to meet today's global challenges. *360° Magazine, Issue 65.*

Stevens, S.S. (1961). To honor Fechner and repeal his law: A power function, not a log function, describes the operating characteristic of a sensory system. *Science, 133(3446),* 80–86.

Stokols, D. (1972). On the distinction between density and crowding: Some implications for future research. *Psychological Review, 79(3),* 275–277.

Stone, J. & Luchetti, R. (1985). Your office is where you are. *Harvard Business Review, 63(2),* 102–117.

Sullivan, L.H. (1896). The tall office building artistically considered. *Lippincott's Magazine 57, March.*

Tanabe, S.-I., Nishihara, N. & Haneda, M. (2007). Indoor temperature, productivity, and fatigue in office tasks. *HVAC&R Research, 13(4),* 623–633.

Taylor, F.W. (1911). *The Principles of Scientific Management.* New York: Harper & Brothers.

Taylor, N.A.S. (2006). Ethnic differences in thermoregulation: Genotypic versus phenotypic heat adaptation. *Journal of Thermal Biology, 31(1),* 90–104.

Thompson, B. (2008). *Workplace Design and Productivity: Are They Inextricably Linked?* London: Royal Institution of Chartered Surveyors.

Tucker, M., Wilson, H., Oseland, N.A, Brogan, P. & Horsley, A. (2020). Unravelling the variables to calculate an organisations' return on workplace investment: A scoping review process. *Proceedings of the Transdisciplinary Workplace Research (TWR) Conference 2020: Future Workplaces.*

Usher, K. (2020). *Your Second Phase: Reclaiming Work and Relationships During and after Menopause.* London: LID Publishing Ltd.

Usher, N. (2018a). *The Elemental Workplace: The 12 Elements for Creating a Fantastic Workplace for Everyone.* London: LID Publishing Ltd.

Usher, N. (2018b). All the workstyles we have ever loved. Retrieved from workessence.com.

Valančius, R. & Jurelionis, A. (2013). Influence of indoor air temperature variation on office work performance. *Journal of Environmental Engineering and Landscape Management, 21(1),* 19–25.

van Meel, J. (2000). *The European Office: Office Design and National Context.* Rotterdam: 010 Publishers.

van Meel, J. (2011). The origins of new ways of working. *Facilities, 29(9),* 357–367.

van Meel, J. (2020). *The Activity-Based Working Practice Guide (2nd Edition).* Amsterdam: Idea Books.

van Meel, J., Martens, Y. & van Ree, H.J. (2010). *Planning Office Spaces: A Practical Guide for Managers and Designers.* London: Laurence King Publishing.

Veldhoen + Company (2018). Activity based working. Retrieved from https://www.veldhoencompany.com/en/activity-based-working/.

Vischer, J.C. (2004). *Designing the Work Environment for Worker Health and Productivity.* Montreal: International Academy for Design and Health.

Vischer, J.C. (2005). *Space Meets Status: Designing Workplace Performance.* London: Routledge.

Vitruvius – Marcus Vitruvius Pollio (30-15BC). *De Architectura*. On architecture, published as *Ten Books on Architecture*, edited and translated into English by F. Granger. Cambridge: Harvard University Press, 1931–1934.

Wargocki, P. & Wyon, D.P. (2017). Ten questions concerning thermal and indoor air quality effects on the performance of office work and schoolwork. *Building and Environment*, 112, 359–366.

Weinberg, A. & Doyle, N. (2017). *Psychology at Work: Improving Wellbeing and Productivity in the Workplace*. London: British Psychological Society.

Weschler, C.J. & Shields, H.C. (1997). Potential reactions among indoor pollutants. *Atmospheric Environment*, 31(21), 3487–3495.

Wilson, E.O. (1984). *Biophilia*. Cambridge: Harvard University Press.

Wolverton, B. C., Douglas, W.L. & Bounds, K. (1989). *A Study of Interior Landscape Plants for Indoor Air Pollution Abatement*, TM-108061. NASA Technical Memorandum.

Wyon, D.P., Anderson, G. & Lundqvist, G.R. (1979). The effect of moderate heat stress on mental performance. *Scandinavian Journal of Work Environment and Health*, 5, 352–361.

Yang, B., Ding, X., Wang, F. & Li, A. (2021). A review of intensified conditioning of personal micro-environments: Moving closer to the human body. *Energy & Built Environment*, 2(3), 260–270.

Yerkes, R.M. & Dodson, J.D. (1908). The relation of strength of stimulus to rapidity of habit-formation. *Journal of Comparative Neurology and Psychology*, 18, 459–482.

# Index

Note: **Bold** page numbers refer to tables; *italic* page numbers refer to figures and page numbers followed by "n" denote endnotes.

absenteeism 77–78, 102
acoustic etiquette 152
acoustic layer and privacy 148–150, *149*
acoustics and noise: solutions for 147–153, **153–154**; sound *vs.* 147; standards of **146**, 146–147
*Action Office 74*, 75, 90, 91, 184, 185
activity-based working (ABW) 76, 132, 135, 164
activity layer 150
activity time 32–33
adaptive behaviour 162, 166; effect on optimum temperature 158, *159*
adaptive comfort 161–163
African Savannah 47, *47*, 55, 102, 103, 183
age 68, 69; physiology 5
Agile Landscaped Office (ALO) 132, 184
agile working 130, 132, 135, 186; adoption of 142–144, *143*; advantages of 137–139, *138*; challenges of 139–141; common terms and strategies 135–137; components of *133*, 133–134; environments 133–137; pre-pandemic adopters of 142
air-conditioned offices 161–163
air movement 102–103
*Allen Curve* 84
Allen, J. G. et al. (2016) 168
Allen, T. 83–84
Allport, G. W. 57–58

Altman, I. 45, 46, 96
ambiverts 178n1
American Society of Heating, Refrigerating and Air-Conditioning Engineers (ASHRAE) 28, 158; indicators of increased performance 28, **29**
*The Anatomy of Human Destructiveness* (Fromm, E.) 48
Andrew, P. 122
animal affinity 105
anthropology 51–52, 183
architectural determinism 71n1
architecture, vernacular 161
Attema, J.E. 171
Attention Restoration Theory (ART) 48

banquette seating 121
Barker, R.B. 99
Barker, R. 42
Barnaby, J.F. 171
Barney, T. 65
Bartlett, P. 19, 29, 36
base zone 98–99
behaviour 12; direct and primary effect on 40; observation of 18
behavioural layer 152–153, **153–154**
Bernstein, E.S. 79–81
The Big Five Personality Inventory 59–61, **61**, 63
biomimicry 49
biophilia 47–49, 101, 183; physiological benefits of 49

*Biophilia* (Wilson, E.O.) 48
biophilic design 101, 188; attachment 113, 114; attributes of 107, *107*; colour 110–111; landscaped office 129; light 111; multisensory 113; patina of time 114; patterns 109–110, *110*; principles of 107, *107*, 108; prospect and refuge 112, *112*, 113; variability 111–112; water 111
*Biophilic Design: The Theory, Science and Practice of Bringing Buildings to Life* (Kellert) 107
booth 117
Bordass, B. 163
Borisuit, A. 174
Bowers, A.R. 171
Boyce, P.R. 173
brainstorm area 123
breakout space 96, 97, 123–124
Brem, A. 81
Briggs, K. 59
Bring Your Own Device (BYOD) 166, 178n2
British Council for Offices (BCO) 22, 24, 25, 36
British Psychological Society 66
Broadbent, D. 63
Broady, M. 40
Browning, B. 101, 115; *The Economics of Biophilia* 49; *14 Patterns of Biophilic Design* 49
Browning, W. 108, 109, 111, 112, 113
Bryan, J.: *Leading People in Change: A Practical Guide* 188
Building Management System (BMS) 164, 165
buildings, forgiveness factor in 163
building standards 114–115
*Bürolandschaft* 74, 74–75, 89–91, 100n1, 100n2, 129, 184, 185
Burt, K. 26
business costs 22, *23*

caddies 140
Cain, S.: *Quiet: The Power of Introverts in a World That Can't Stop Talking* 62
Calabrese, E. 107–114
Calhoun, J. B. 12
calming/stimulating base zone 98, 99, 151, *151*, 152

carbon dioxide ($CO_2$) 166–168
care 94–95, 123–124
carrel 121
Carter, D. 173
Cattell, R.B. 58
Centraal Beheer 75
Chiat/Day 132, 144n1
chillout area 124
choice 92, 134–137, 151, 186; and control 152
circadian rhythm 175, *175*
Clancy, J. 108, 109, 110, 111, 113
Claudio, L. 105
climate chambers 160
club 132
co-creation 95
collaboration 92–93, 121–123, *122*
co-location 95
colour 39, 45, 67–69, 80, 92, 93, 99, 109–111, 115n1, 129, 148, 160, 164, 172, 174–176
combi-office 75
comfort 93–94
common spaces 124
concentration 93, 120–121, *120*
concrete jungle 4
confidentiality 94, 120–121, *120*
connectivity 94, 121–123, *122*
contemplation 94, 105, 123–124
contextual factors 160
contiguous single space 125
continuity 139
control 94
Cool Biz 162
core 124
Corporate Real Estate (CRE) industry 14n1, 17, 18, 37, 54, 130, 181; biophilic design 101; focus of 19
cortisol 175, 176
cost 95, 181, 186; balancing performance and 19, *19*; employee 22; and performance relationship on productivity 19, *20*; reducing property 130
Costa, P.T. 60
cost-benefit analysis 21, 34–36, 182
cost-cutting 35–36
Covid-19 pandemic 11–12, 27, 44, 142, 143, 185
co-working spaces 25

creativity 93, 121–123, *122*
crowding *vs.* density 45
C-suite 92, 100n3
culture 70

Dahl, R. 144n2
DARE principles 147–148
Dart, R. 71n3
Darwin's theory of evolution 47
Dawkins, R.: *The Selfish Gene* 7
daylight 39, 45, 102, 172; and colour
    spectrum 174–176, *175*
"DDA compliance" 69
de Dear, R. 78
decision-making performance 168
DEGW 10; "Three Es" of 19
Democratic User Control of Zonal
    Temperature (DUCOZT) 165
demographic factors 5, 9, 182
density: *vs.* crowding 45; poor open
    plan 89–90, *90*; workplace 149
design: and ambience 99; biophilic
    49; clothing and building 161;
    collaboration 92–93; of furniture
    116–117; multisensory 64–66
desk: alternative configurations
    118–119, *119*; non-rectangular and
    non-orthogonal 126, *127*; standard
    configurations 117, *118*
desk clusters 96–97, *97*, 98
desk screens 149, *149*
desk sharing 25–27, 37n5, 142
desk size 25, 99
desk utilisation rates 26
Despenic, M. 173, 174
destination density 24–27
dialectic 45
*Dictionary of Construction, Surveying,
    and Civil Engineering* (Gorse,
    Johnston and Pritchard) 90
dimensions of extroversion and
    neuroticism 60
disabilities 69–70
Disability Discrimination
    Act 1995 69
diversity 66–71
Duffy, F. 100n1, 132
Dunbar, R. 51, 52
Dunbar's Number 51–52, 98, 183
Dunn, W.: model of sensory processing
    65–66

Eary, J. 186
ecological valence theory (EVT) 110
ecological validity 31
*The Economics of Biophilia*
    (Browning) 49
efficiency 8, 10, 18, 19, 21, 22, 24, 25,
    27, 72–74, 82, 89, 96–98, 118, 124,
    130, 136, 139, 149, 186
85:15 staff: property ratio 22
*Elemental Change* (Usher) 188
*The Elemental Workplace* (Usher) 70
Eley, J. 117
enclosed spaces 149
enticement 139
environment 4, 6, 7, 8, 9, 11, 32, 38,
    40, 41–48, 50, 56, 59, 65, 66–68, 77,
    80, 81, 82, 90, 93, 99, 102, 105, 108,
    109, 111, 113, 116, 124, 125, 127,
    129, 132, 137–140, 142, 151, 153,
    158, 160, 163, 174, 182, 184, 187
environmental conditions 186
environmentally responsive
    workstation 165, *165*
Environmentally Responsive
    Workstation (ERW) 165, 166
environmental psychologists 31
environmental psychology 8, 41–42,
    182–183; spaces for people
    43–46, *44*, *45*
Equality Act 2010 69
Erasmus, A.W. 90
ethnic groups 7
European Association of Zoos and
    Aquaria (EAZA) 18
European economic crisis 74
Eve, G. 24
evidence-based design 8–11, *9*
evolution 46–48, *46*, 50, 102, 132
evolutionary psychology 46, 46–47, *47*,
    102, 183; air movement 102–103;
    animal affinity 105; anthropology
    and Dunbar's number 51–52;
    biophilia 47–49; contemplative 105;
    daylight 102; greenery 105, 107, *106*;
    inquisitive 103; movement 103–104;
    neuroscience 50, *50*–51; socialising
    104, *104*; sound 103; temperature
    102; views 102
expectation 42, 64, 140, 153, 182
experience 9, 40, 42, 49, 56–58, 60, 63,
    65, 66, 108, 134, 156, 157

extroversion: dimensions of 60;
    personality dimensions 58
extroverts 56–58, 61–64, 67; acoustic-
    related solutions for 153, **153–154**;
    arousal of 63
Eysenck, H.J. 56, 58; dimensions of
    extroversion and neuroticism 60
Eysenck, S.B.G. 56

facilities management (FM) 20
Fechner, G. T. 38
Ferreira, E.J. 90
*firmitas* 9, 10
Fisk, W. 155, 168
Five Factor Model *see* The Big Five
    Personality Inventory
flexible working 132, 135, 141
flourish 5, 7, 13, 37, 43, 48, 182
focus/quiet space *120*, 121
"forest bathing" 49
"form follows function" 9
*14 Patterns of Biophilic Design*
    (Browning) 49
free-range office 76, *76*, 91, 184
Freud, S. 57
Fromm, E. 101; *The Anatomy of Human
    Destructiveness* 48
functional comfort 53

Galen 56, 58
Gates, B. 136
gender 1, 5, 44, 69, 158, 177, 183
generation 68–69
Gerba, C. 12
glare 177
Goldsmith, S. 69
Gore 52, 71n4
greenery 105, 107, *106*
Groenewald, D. 90

Habitability Pyramid 53, 54
Hall, E.T. 99; *Proxemic
    Framework* 43, 99
Hamilton, D. K. 8
Haneda, M. 156
Harris, R. 72, 73
health crisis 11–12
health, physiology 5
Hediger, H. 43
Heerwagen, J. 5

Herczeg, L. 168
Herzberg, F. 30, 54–55, 55, 93, 183
Heschong, L. 48
heterogeneous groups 62
high-density 7, 25, 44, 45, 72, 75, 78,
    80, 82, 89, 90, 130, 185, 188
higher-order needs 53
Hippocrates 56, 58
Hiroshi, S. 171
Hodsman, P. 64, 65, 145, 147
Holy Grail 17, 18, 37n1
home 93, 97, 134, 136, 143, 160
homo sapiens 46
Hongisto, V. 82
hormones 39, 175
human-centric approach 8–11, *9*, 92
*The Human Zoo* (Morris) 4
humidity 112, 157
hybrid working 136–137, 186, 188
hygiene factors 54

inclusivity 183–184; and diversity 66–71
indoor air quality: IAQ and
    performance 167–169, *169*;
    regulation 166–167; scents and
    odours 169–170; solutions for 170
indoor environmental conditions
    145, 177–178; acoustics and noise
    146–154; indoor air quality 166–170;
    lighting and daylight 170–177;
    thermal comfort 155–166
informal meeting areas 123
in-house business metrics 31–32
inquisitive 103
Institute of Workplace and Facilities
    Management (IWFM) 33, 35
interaction, social 104
interior design 14, 95, 112, 186
introversion 147, 178n1
introverts 56–58, 60–64; acoustic-
    related solutions for 153, **153–154**;
    arousal of 63
intuition function 58
*ISO Standard 22955: Acoustic Quality of
    Open Office Spaces* 146
*It's a Zoo Around Here* (Risner) 7

Japan: clothing and building design
    161; Cool Biz 162
Johnson, A. 144n2

*Journal of Environmental Psychology* 41
Jung, C. 57
Jurelionis, A. 156

Kane, C. 10, 17, 20, 21
Kaplan, R. 48
Kaplan, S. 48
Katjár, L. 168
Kellert, S. 48, 101, 107–115; *Biophilic Design: The Theory, Science and Practice of Bringing Buildings to Life* 107
Kim, J. 78
Kim, S. 174
King Louis XIV 161
Knight, C. 49
Knowles, M. 77
Kotter, J.P. 188

landscape office 13–14, 89, 184–187, 189; accommodating human needs 95–99, *97, 98*; (re)emerging workplace solution 89–92, *90, 91*; example floor plate *128*, 128–129; humanising office 92–95; implementation of 187–188; look and feel 129; physical components to 184, *184*; planning of 124–127; reckoning occupational density 130; traditional workplace layout to 187–188; workplace layout 124–125; work-settings in 116, 126–127, *127*; workspace zones 125–126, *126*; *see also Bürolandschaft*
Lan, L. 155, 156
*The Lawyer* 27
layout: office *74,* 74–76; workplace 125; workspace 185
*Leading People in Change: A Practical Guide* (Bryan) 188
Leaman, A. 163
Le Corbusier 8, 144n2
Lee, J.-H. 174
Lee, P.D. 54
Leesman Index 18, 21, 80, 85n2, 137
legacy workplace issues 11–12
Lei, Q. 155
Lewin, K. 41–42
Lian, Z. 155

life stage 68–69
lighting 67, 68, 111; and performance 170–171, *171*; and preference *173*, 173–174; solutions for 176–177; standards of 171–172
limbic system 50, 51
Lister, K. 54
Liverpool John Moores University (LJMU) 29, 33, 35
Loftness, V.E. 30
London 73
London Underground 6
loneliness 83
lower-order needs 52–53
Luchetti, R. 76, 132

Mace, R. 70
MacLean, P. D. 50
Malmberg, T. 140
Malnar, J. M.: *Sensory Design* 65
Marini 144n4
Marmot, A. 117
Marsh, M. 65
Maslin, S. 67
Maslow, A.H. 52–54, *53,* 71n5, 183
Maslow's hierarchy of needs 52–54, *53,* 94
Maula, H. 168
Mayo, E. 42
McCrae, R.R. 60
McLaurin, J. P. 78–79
measurement 32; of performance 27–34; of productivity 17; of relationship between physical stimuli and sensory response 38, *39*
meat locker effect 160
Meek, C. 171
meeting rooms 27, 62, 63, 92, 98, 99, 110, 122, 123, 134, 135, 167, 168, 177
meeting space 9, 67, 76, 80, 85n1, 92, 95, 104, 117, 121, 122, *122,* 126, 127, 134, 141, 147, 148, 150, **153, 154,** 183
Mellon, C. 167–168
Miller, H. 75, 76, 85n1
Mixson, E. 81, 82
modern office: air-conditioned glass towers 161; history of 72

modern open plan offices 72, 89, 91, 97, 149
modern zoo design 5, 6, *6*
monomyths 37n1
Moon, J.W. 174
Moore, T. 173
Morrell, P. 17
Morris, D. 103, 105; *The Human Zoo* 4; *The Naked Ape* 3
Motivation-Hygiene Theory *see* Two-Factor Theory
motivation theory 183; Herzberg 54–55, *55*; Maslow 52–54, *53*
movement 64, 65, 68, 97, 103–104, 111, 125, 126, 161
Mueller, K. 65
Müller-Lyer illusion 40, *40*, 43, 95
multimedia hub 123
multisensory 13, 64–66, 113, 114, 183
Myers-Briggs Type Indicator (MBTI) 59, *59*
Myers, I. 59
Myerson, J. 68

*The Naked Ape* (Morris) 3
National Bureau of Standards 22
National Electrical Manufacturers Association (NEMA) 28; indicators of increased performance 28, **28**
nature 47–49, 95, 96, 101–103, 107–111, 113, 114, 119, 129
Nelson, G. 75
Net Internal Area (NIA) 24, 37n4, 130
neurodiversity 67–68
neuroscience 50, 50–51
neuroticism: dimensions of 60; personality dimensions 58
Newsham, G. 173
new ways of working (NWW) 136
19th-century zoo 4, *4*
Nishihara, N. 156
noise *see* acoustics and noise
noise distraction 64
Noise Rating (NR) curves 146
"non-taxing involuntary attention" 48
nooks and crannies 92, 103, 125, 183

occupational density 130, 131n1, 181
OCEAN 60
Odbert, H.S. 57–58
odours 169–170

office design 13, 182; *see also* design
office landscape 90
office layout 74, 74–76
office occupants 42, 84, 119 , 157, 158, 167, 178, 185–187
office simulations 43
office spaces 72; types of 103–105, 107, *104*, *106*
Ohly, H. 48
Oldman, T. 21
1:1 space 121
Oommen, V.G. 77
open area 117
open plan desking 82
open plan office 91; acoustic parameters for 146, **146**; benefits of 81–85; desk/surface illuminance for 172; interpretation of 182; *vs.* private offices 77–81; route to 72–76, *73*, *74*, *76*; workspace layout 185
operative temperature 157
optical illusions 40, *40*
optimal occupational density 91
optimal workspace solution 90
optimum temperature, adaptive behaviour effect on 158, *159*
organic 74, 75, 91, *127*, 185
orthogonal 75, 96, 129
Oseland, N. 138
Oseland, N.A. 29, 36
Osmond, H. 43, 96
Oxley, D. 43

paired/shared offices 84, 118
Palmer, S. 110
pandemic 11–12, 27, 44
Parker, C. 145
partition 75, 80, 89, 90, *90*, 96, 116, 122, 149
patina of time 114
patterns 109–110, *110*
Pejtersen, J. 77–78
Pell, N. 76
people-centric approach 21
perceived space, factors affecting *44*, 44–45
percentage of people dissatisfied (PPD) 178
performance 18, 138; balancing cost and *19*; and cost relationship on productivity 19, *20*; factors affecting

*30*, 30, *31*; IAQ and 167–169, *169*; illuminance effect on 171, *171*; in-house indicators of 29; lighting and 170–171, *171*; measuring of 27–34; temperature and *155*, 155–156; ventilation rate effect on 168, *169*

performance metric: using feedback as 34; workplace impact on 33, **33**

personal benefits 138

personal factors 5

personalisation 139–140

Personalised Environment Module/ Climadesk 165

personality 5, 7, 9, 13, 38, 40, 42, 44, 55–64, 67, 71, 84, 94, 96, 98, 118, 120, 140, 147, 150, 151, 152, 177, 178, 183, 186

personality theory 55–56, 183; temperaments and psychoanalysis 56–57, *57*; trait theory 57–64, *58*, *59*, **61**, *63*

personality traits 5, 44, 58, 60, 151

personal space 25, 43–45, 97, 99, 127, 144, 149

phone booths 121

physiology 1, 5, 13, 38, 46, 68, 102, 156, 174, 178, 186

planting 93, 101, 102, *104*, 109, 114, 129

pod 26, 80, 92, 93, 96, 116–117, 121–125, 127, 129, 130, 134, 149, 154, 186

Poisson regression 77–78

Pollard, E.L. 54

poor open plan 90, *90*

Portable Appliance Testing (PAT) 166, 178n3

post-occupancy evaluations (POEs) 6, 36, 145

post-pandemic 13, 186, 188; *see also* Covid-19 pandemic

Predicted Mean Vote (PMV) method 157–158

preference *173*, 173–174

preferred: environmental conditions 103, 182; interpersonal distance 44; meeting space 67; temperature 160, 164, 165; workspaces 5

presenteeism 12

primitive/survival brain *see* reptilian complex

privacy 75, 97; factors affecting *44*, 44–45

privacy regulation theory 45, *45*, 96

private offices 84; open plan *vs.* 77–81; preference for 83; standard desk and 117, *118*

productivity 37n2; in business 18–19; measuring of 17; staff costs 22

productivity equation 18–20, *19*, *20*

project evaluation 36–37

project room 123

property costs 22, 24, *23*

Propst, R. 75

prospect 112, *112*

*Proxemic Framework* (Hall) 43, 99

psychoanalysis 56–57, *57*

psychological comfort 53–54

psychology 8, 9, 13, 38, 41–52, 57, 62, 66, 71, 92, 97, 102–107, 110, 129, 178, 182, 183, 186, 187

psychophysicists 38, 39, 41

psychophysics 38–41, *39*, *40*, 182

Quickborner consulting group 74, 75

*Quiet: The Power of Introverts in a World That Can't Stop Talking* (Cain) 62

quiet zone 93, 96, 121, 124, 129, 134, 148, 152

Raisbeck, K. 54

recruitment agencies 69

rectangularity illusion 43

refuge 112, 113, *113*

reliability 30

remote working 22, 35, 69, 95, 133–136, 139, 152

reptilian complex 50–51

revenue per square metre 27

Risner, N.: *It's a Zoo Around Here* 7

Ritter, S. 48

Rohles, F. 160

Rothe, P. 81

Royal Institute of British Architects (RIBA) 36

Ryan, C. 108, 109, 111, 113

Sadalla, E. 43

Saint Augustine 72, *73*

Saint Jerome 72, *73*

Sargent, K. 66, 67
Satish, U. 168
Savannah Hypothesis 46, 47, 71n3, 103
SBS *see* Sick Building Syndrome (SBS)
scents 169–170
Scheiberg, S. L. (1990) 140
Schloss, K. 110
Schnelle, E. 74
Schnelle, W. 74
Scientific Management 73
screen-based activity 176–177
screens 149, *149*
*The Selfish Gene* (Dawkins) 7
self-reported performance 29, 34, 77–78
"sensation avoiding" 66
"sensation seeking" 66
senses 64–65
sensing function 58
sensitivity 65
*Sensory Design* (Malnar and Vodvarka) 65
Sensory Intelligence 66
sensory processing 64–66
Seppänen, O.A. 155, 168
shared workspaces 134, 139
Sharp 145
Shaw, G. B. 144n2
Shinrin-Yoku 49
Sick Building Syndrome (SBS) 11, 167
signals and visual cues 152–153
Signature Aromas 169
situational factors 5, 44
Sixteen Personality Factors (16PF) model 58
Slater, A. 173
smart working 136
social democratic office 75
social interaction 104
socialising 104, *104*
sociofugal space 43, 96, 97
sociology 13
sociopetal space 43, 96, 97, *97*
Sommer, R. 43
sound 103; *vs.* noise 147
Sound Pressure Level (SPL) 146
Sowa, J. 163
space: efficiency 19, 130; for people 43–46, *44*, *45*
space plan 124, 125, 128–130

species 4–7, 46–48, 53, 55, 71, 164
Speech Intelligibility Index (SII) 147
Speech Transmission Index (STI) 147
standard desk and private offices 117, *118*
Stangl, K.: *The Workplace Zoo* 7
Steelcase 44, 123
Stevens's Power Law 39
Stevens, S. S. 39
Stewart, N. 171
stimulus-response relationships 41
Stokols, D. 45
stress hormone 175
Sullivan, L. H. 9
Sundstrom, E. 139
super-trait theory 56
survive 13, 37, 46–47, *47*, 55, 102, 110, 183
sustainability 138, 139
Swanke Hayden Connell Architects 76

Takasago Corporation 169
Tanabe, S.-I. 156
task lighting 172
task performance, design element impact on 33, **33**
Taylorism 28, 73
temperaments 56–57, *57*
temperature 102; for animals 5, 6; and performance 155, 155–156
theory of personality 58, *58*
thermal comfort: adaptive comfort 161–163; solutions for 163–166, *165*; standards of 156–158, *159*, 160; temperature and performance 155, 155–156
thermal zones 164–165
thinking function 59
"Three Es" 10, 19
Three-Factor Model for workplace 55, *55*
thrive 5, 13, 37, 55, 94, 101, 102, 151, 182
touchdown space 121
trait theory 57–64, *58*, *59*, **61**, 63
Triune Brain system 50, *50*, 51
Tucker, M. 29
Turban, S. 79–81
two-dimensional horizontal space 92
Two-Factor Theory 54–55, *55*, 93

UK: occupational density in 24; open plan offices 84
United States, for cubicle design 75
Usher, K. 69
Usher, N. 89, 92; *Elemental Change* 188; *The Elemental Workplace* 70
*utilitas* 9, 10

Valančius, R. 156
validity 31
value 21, 36, 37n3
value-cost conundrum 20–22; destination density 24–27; property costs 22–24, *23*; quantifying benefits 27–34
Van der, V. 140
van Meel, J. 72, 132, 140
variability 111–112
Veldhoen + Company 76, 135
*venustas* 9, 10
vernacular architecture 161
views 102
Vischer, J. 53
Vitamin D 176
Vitruvian architectural principles 9, *9*
Vodvarka, F.: *Sensory Design* 65
Volatile Organic Compounds (VOCs) 166–168
de Vries, A. 171

Wargocki, P. 155, 168
water 111
Watkins, D. 8
Webber, C. 138
Weber, E. H. 38
Weber–Fechner law 39
wellbeing 3, 5–9, 13, 21, 25, 29, 37, 38, 46, 49, 51, 53, 54, 56, 64, 66, 71, 89, 92, 95, 101, 132, 135, 137, 138, 140, 141, 144, 145, 151, 176, 177, 181–189
*WELL Building Standard (WELL)* 36, 37n6, 114–115, 146, 157, 164
well-designed office lighting 172
Wells, W. 160
Western Electric Company 34, *35*, 42
WGBC *see* World Green Building Council (WGBC)
Wilson, E. O. 101; *Biophilia* 48

work 3
working from home (WFH) 72, 133, 140
work patterns 140–141
workplace 101; desk sharing 25–27; differences between different strategies 91, **91**; generations in 68–69, 71n6; impact on performance metrics 33, **33**; multisensory design 64–66; overcrowding in 25; on performance dates 28; on productivity 18; Three-Factor Model for 55, *55*
workplace design, human needs and impact on 182
workplace industry 14n1, 70, 95, 99, 135
workplace layout, landscape office 124–125
"workplace taboo" 69
*The Workplace Zoo* (Stangl) 7–8
workplace zoo, origins of 3–8, *4*, *6*
work-settings 185; collaboration, creativity and connectivity 121–123, *122*; concentration and confidentiality 120–121, *120*; contemplation and care 123–124; core and common spaces 124; design and structure of furniture 116–117; in landscaped office 116; landscape office 126–127, *127*; location of 150; menagerie of 119–124
workspace 72; and associated cost 19; open plan (*see* open plan offices)
workspace zones, landscape office 125–126, *126*
World Green Building Council (WGBC) 22, 29, 166, 167
Wright, F. L. 74
Wyon, D.P. 155, 168

Yang, B. 165
Yerkes-Dodson Law 62
Young, D. 101

Zhao, I. 77
zoning layer 150–152, *151*
zoo enclosure design 5, 7, 53
zoology 13